Synthesis Lectures on Renewable Energy Technologies

The series, Synthesis Lectures on Renewable Energy Technologies publishes concise books, focused on technologies that harness energy from naturally occurring sources, such as sunlight, wind, water, geothermal heat, and biofuels from organic materials. These renewable energy technologies play a crucial role in transitioning away from fossil fuels, helping to mitigate the effects of climate change, and promoting a sustainable energy supply.

Hemantkumar N. Akolkar · A. K. Haghi ·
Nirmala R. Darekar · Ganesh B. Pawar

Bioenergy from Microalgal Biomass

Sustainable Conversion Technologies

Springer

Hemantkumar N. Akolkar
Department of Chemistry
Abasaheb Marathe College
Rajapur, Maharashtra, India

Nirmala R. Darekar
Department of Chemistry
Radhabai Kale Mahila Mahavidyalaya
Ahmednagar, Maharashtra, India

A. K. Haghi
Edinburgh, UK

Ganesh B. Pawar
Department of Botany
Abasaheb Marathe College
Rajapur, Maharashtra, India

ISSN 2690-5000 ISSN 2690-5019 (electronic)
Synthesis Lectures on Renewable Energy Technologies
ISBN 978-3-032-03433-5 ISBN 978-3-032-03434-2 (eBook)
https://doi.org/10.1007/978-3-032-03434-2

This Springer imprint is published by the registered company Springer Nature Switzerland AG
The registered company address is: Gewerbestrasse 11, 6330 Cham, Switzerland

Preface

Sustainable and renewable energy sources are becoming more and more necessary as the globe deals with an extraordinary energy crisis. Bioenergy production feedstocks derived from microalgal biomass have recently gained attention as a viable, environmentally friendly, and efficient replacement for conventional fossil fuels. This book, *Bioenergy from Microalgal Biomass: Sustainable Conversion Technologies* offers a thorough examination of recent developments in microalgal bioenergy production, encompassing biomass cultivation, harvesting, conversion technologies, and sustainability aspects.

The book comprises five chapters, each focusing on a significant aspect of microalgal bioenergy production. Chapter 1 presents the principles of green chemistry and bioenergy, emphasizing the properties and benefits of microalgal biomass. Chapter 2 examines the production and harvesting of microalgal biomass, addressing microalgal strains, cultivation techniques, photobioreactor systems, and biomass pre-treatment methods. Chapter 3 examines several biomass conversion processes, such as thermochemical conversion, biochemical conversion, hydrothermal liquefaction, and biogas production via anaerobic digestion. Chapter 4 examines the production of biodiesel and bioethanol, focusing on the extraction and processing of microalgal biodiesel, transesterification reactions, and fermentation processes. Chapter 5 analyses integrated biorefinery concepts, the design of algal biofilms and photobioreactors, as well as the application of genomics and metabolic engineering in microalgal bioenergy. This chapter examines the challenges and opportunities associated with microalgal bioenergy, emphasizing future trends and directions.

This book targets researchers, professionals, and students engaged in bioenergy, renewable energy, and sustainable technologies. This resource is essential for understanding recent advancements and applications in microalgal bioenergy production. This book, featuring contributions from field experts, seeks to inspire additional research and innovation, thereby promoting a sustainable and environmentally friendly energy future.

Rajapur, India — Dr. Hemantkumar N. Akolkar
Ahmednagar, India — Dr. Nirmala R. Darekar
Rajapur, India — Dr. Ganesh B. Pawar
Edinburgh, UK — Dr. A. K. Haghi

Contents

About the Authors

Hemantkumar N. Akolkar Ph.D., is currently working as an Associate Professor in the Department of Chemistry at Rayat Shikshan Sanstha's Abasaheb Marathe Arts and New Commerce Science College, Maharashtra, India. He is an academician and researcher who has worked for the last 14 years at both undergraduate and postgraduate levels. He has published five academic research-oriented books and 60+ research papers in national and international journals and conferences of repute. His areas of interest are heterocyclic chemistry, synthetic organic chemistry, and green chemistry. He is a reviewer of several journals of international repute in chemistry.

A. K. Haghi Ph.D., is a retired Professor of Engineering Sciences and has published 250+ academic research-oriented books and 1000+ research papers in various journals and conference proceedings. His leadership in academic publishing includes founding and serving as Editor-in-Chief of prestigious journals. Professor Haghi's extensive educational background and supervisory roles underscore his expertise and contributions to the field of engineering sciences. He is currently appointed as Honorary Research Associate (HRA) at University of Coimbra, Portugal.

Nirmala R. Darekar Ph.D., is currently working as an Assistant Professor in the Department of Chemistry, Rayat Shikshan Sanstha's Radhabai Kale Mahila Mahavidyalaya, Ahmednagar, Maharashtra, India. Her specialization is in Analytical Chemistry. She has 12+ years of experience in teaching and six years of experience in research. She did her Ph.D. from Savitribai Phule Pune University, Pune, India. She has published two books, 13 chapters, three patents and more than 16 Research papers in reputed International journals.

Ganesh B. Pawar Ph.D., is currently working as an Assistant Professor in the Department of Botany at Rayat Shikshan Sanstha's Abasaheb Marathe Arts and New Commerce Science College, Maharashtra, India. He has 12+ years of experience in teaching and five years of experience in research. He did his Ph.D. from Shivaji University, Kolhapur, India. He has published two books, one patent and 10+ Research papers in reputed journals.

List of Figures

List of Tables

Introduction and Fundamentals of Green Chemistry and Bioenergy

1

Abstract

The increasing worldwide need for energy, along with concerns about climate change and environmental sustainability, has reignited interest in alternative energy sources. Bioenergy derived from renewable biomass has emerged as a promising answer. However, traditional bioenergy production methods usually use non-sustainable practices, causing environmental degradation and resource depletion. Green chemistry lays the groundwork for developing long-term bioenergy production techniques that are both environmentally friendly and energy efficient.

Keywords

Green chemistry • Microalgae • Bioenergy

1.1 Introduction to Green Chemistry and Bioenergy

1.1.1 Definition and Principles of Green Chemistry

Green chemistry, also known as sustainable chemistry, is a strategy of developing chemical products and processes that reduces or eliminates the use and production of toxic substances. Paul Anastas and John Warner proposed the 12 principles of green chemistry in 1998 [1]. These principles are intended to lessen the environmental effect of chemical goods and processes:

H. N. Akolkar et al., *Bioenergy from Microalgal Biomass*, Synthesis Lectures on Renewable Energy Technologies, https://doi.org/10.1007/978-3-032-03434-2_1

1. **Prevention**: Rather of treating or cleaning up waste, prevent it from generating in the first place.
2. **Atom Economy**: Synthesize materials using methods that maximize the incorporation of all starting materials.
3. **Less Hazardous Chemical Syntheses**: Use safer, less toxic substances and processes.
4. **Designing Safer Chemicals**: Design products that minimize harm to human health and environment.
5. **Safer Solvents and Auxiliaries**: Use innocuous, easily separable, and recyclable solvents.
6. **Design for Energy Efficiency**: Minimize energy consumption and optimize energy efficiency.
7. **Use of Renewable Feedstocks**: Use renewable resources instead of depleting non-renewable ones.
8. **Reduce Derivatives**: Minimize unnecessary chemical modifications.
9. **Catalysis**: Use catalysts to reduce reaction temperatures, pressures, and chemical waste.
10. **Design for Degradation**: Design products that break down safely and easily.
11. **Real-Time Analysis for Pollution Prevention**: Monitor and control processes in real-time.
12. **Inherently Safer Chemistry for Accident Prevention**: Design processes that minimize risk.

1.1.2 What Is Bioenergy?

Bioenergy is the energy produced by organic materials, such as plants, wood, and garbage, and is turned into a variety of energy forms, including heat, electricity, and transportation fuel [10]. This renewable energy source provides a sustainable alternative to fossil fuels, lowering greenhouse gas emissions and mitigating climate change [18].

Bioenergy is a type of renewable energy since the energy contained in biomass is sunlight collected through natural photosynthetic processes, and as long as the amount of biomass utilized is equal to or less than the amount that can be regrown, it is potentially renewable indefinitely. Fuels and power produced from biomass are referred to as bioenergy. For instance, electricity produced by burning biomass is known as biopower. By burning biomass, heat, steam, or a combination of the two can also be created in a process known as co-generation, which also produces electricity. The term "biofuel" is frequently used to describe liquid fuels and gasses generated from biomass that are primarily utilized in transportation [20].

1.2 Microalgae: Characteristics and Advantages

Microalgae are photosynthetic microorganisms that are categorized as prokaryotes or eukaryotes. Blue-green algae are prokaryotic microalgae, similar to cyanobacteria, whereas diatoms and green algae are eukaryotic microalgae. Ghasemi et al. [9], Ghasemi et al. [8], Rasoul-Amini et al. [17], Mostafa [16], and Fu et al. [7] found that these microbes may thrive in a wide range of habitats and adapt to mild to severe conditions. According to Zhang et al. [22], they are a varied group of photosynthetic eukaryotes that range from unicellular to multicellular, and they have a considerably simpler cellular structure than other plants. Microalgae consists of three major chemical components: proteins, hydrocarbons, and lipids [14]. Microalgae have a number of economic and environmental benefits over other organisms, such as quick growth, the capacity to develop at different nutritional trophic levels, high oil generation, and high photosynthetic efficiency, which helps with wastewater treatment and CO_2 biomitigation [11]. Additionally, microalgae may be farmed on non-arable land, minimizing competition for land from food crops. Moreover, microalgae have been shown to be among the most durable species in their surroundings. Under stressful situations, physiological changes may occur that modify biomass composition and nutritional content [12]. This significant characteristic of microalgae has been employed in various research to adjust the nutrients in microalgae cultivation in order to create biomass with high quantities of protein, lipid, and carbs for particular applications [2, 21].

1.3 Overview of Bioenergy Production from Microalgae

Sustainable bioenergy production is being recognised as an effective approach to mitigate greenhouse gas emissions and decrease dependence on fossil fuels. Algae represent a distinct category of feedstock utilised in the generation of biofuel. Algae are a desirable feedstock for the generation of gaseous and liquid biofuels without competing with the food production business [6]. Algae are categorized into two types: macroalgae, which consists multicellular seaweeds, and microalgae, which are microscopic, unicellular algae found in a variety of environments and consist of various species with distinct evolutionary backgrounds [19].

Algae are autotrophic organisms that synthesise organic macromolecules like carbohydrates and lipids from solar energy and atmospheric CO_2. These compounds may then be transformed into biofuel. As a consequence, algae may be able to provide a carbon–neutral fuel with minimal inputs. Although they need carbon input, several algae species may also be cultivated heterotrophically to improve yield. There are several situations for using algae to produce biofuel.

All kinds of algae may be utilized as a feedstock for thermochemical conversion processes such hydrothermal liquefaction, gasification, and pyrolysis to produce liquid and

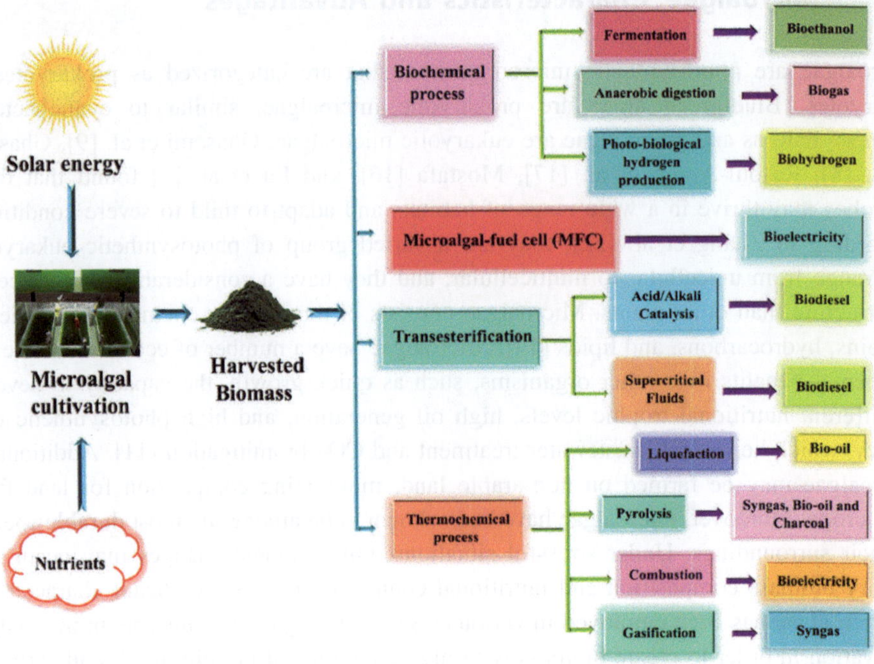

Fig. 1.1 Schematic demonstrating the methods for turning microalgal biomass into various bioenergy forms [4]

gaseous biofuels [13]. Macroalgae are an excellent source of carbohydrates that can be transformed into bioethanol or biomethane by biological conversion processes including fermentation and anaerobic digestion, however they only collect very little amounts of oils. The high oil yields seen in some strains of microalgae, namely the glycerolipid triacylglycerol (TAG), which may be transformed into biodiesel by transesterification, make them particularly interesting [3, 5].

All things considered, the main appeal of algae as a feedstock for biofuel is that it may be grown in open ponds, photobioreactors, or occasionally heterotrophic fermentative conditions on non-agricultural land. Since many strains can grow well in saltwater or wastewater, the cultivation might take place in environments that don't require freshwater [3, 15].

Nowadays, microalgae-based bioenergy is seen as a viable and environmentally responsible substitute for meeting the world's energy demands and reducing pollution. Through various thermochemical, biochemical, or chemical processes, microalgal biomass and the biomolecules that make it up may be converted into a range of biofuels, such as biodiesel, bioalcohol, biogases (methane, hydrogen, syngas, and hythane), and bioelectricity (Fig. 1.1).

References

1. Anastas, P. T., & Warner, J. C. (2000). *Green chemistry: theory and practice.* Oxford university press.
2. Aslam, A., Thomas-Hall, S. R., Manzoor, M., Jabeen, F., Iqbal, M., Uz Zaman, Q., Schenk, P.M., & Tahir, M. A. (2018). Mixed microalgae consortia growth under higher concentration of CO_2 from unfiltered coal fired flue gas: fatty acid profiling and biodiesel production. *Journal of Photochemistry and Photobiology B: Biology, 179*, 126–133. https://doi.org/10.1016/j.jphotobiol.2018.01.003.
3. Brennan, L., & Owende, P. (2010). Biofuels from microalgae—a review of technologies for production, processing, and extractions of biofuels and co-products. *Renewable and sustainable energy reviews, 14*(2), 557–577.
4. Choudhary, S., Tripathi, S., & Poluri, K. M. (2022). Microalgal-based bioenergy: Strategies, prospects, and sustainability. *Energy & Fuels, 36*(24), 14584–14612.
5. Demirbas, A. (2009). Biofuels securing the planet's future energy needs. *Energy conversion and management, 50*(9), 2239–2249.
6. Driver, T., Bajhaiya, A., & Pittman, J. K. (2014). Potential of bioenergy production from microalgae. *Current Sustainable/Renewable Energy Reports, 1*, 94–103. https://doi.org/10.1007/s40518-014-0011-8.
7. Fu, W., Nelson, D. R., Mystikou, A., Daakour, S., & Salehi-Ashtiani, K. (2019). Advances in microalgal research and engineering development. *Current opinion in biotechnology, 59*, 157–164.
8. Ghasemi, Y., Rasoul-Amini, S., Hossein Morowvat, M., Bagher Ghoshoon, M., Javad Raee, M., Khoubani, S., Negintaji N., Nouri F. & Parvizi, R. (2009). C-20 ketone reduction of hydrocortisone by rice field microalga Chlorella vulgaris MCCS 013. *Chemistry of Natural Compounds, 45*, 824–828.
9. Ghasemi, Y., Rasoul-Amini, S., Morowvat, M. H., Raee, M. J., Ghoshoon, M. B., Nouri, F., ... & Mosavi-Azam, S. B. (2008). Characterization of hydrocortisone biometabolites and 18S rRNA gene in Chlamydomonas reinhardtii cultures. *Molecules, 13*(10), 2416–2425.
10. Himmel, M. E., & Bayer, E. A. (2009). Lignocellulose conversion to biofuels: current challenges, global perspectives. *Current Opinion in Biotechnology*, 20(3), 316–317.
11. Hussain, F., Shah, S. Z., Ahmad, H., Abubshait, S. A., Abubshait, H. A., Laref, A., Manikandan, A., Kusuma, H.S. & Iqbal, M. (2021). Microalgae an ecofriendly and sustainable wastewater treatment option: Biomass application in biofuel and bio-fertilizer production. A review. *Renewable and Sustainable Energy Reviews, 137*, 110603. https://doi.org/10.1016/j.rser.2020.110603.
12. Jiang, L., Pei, H., Hu, W., Hou, Q., Han, F., & Nie, C. (2016). Biomass production and nutrient assimilation by a novel microalga, Monoraphidium spp. SDEC-17, cultivated in a high-ammonia wastewater. *Energy Conversion and Management, 123*, 423–430. https://doi.org/10.1016/j.enconman.2016.06.060.
13. Jin, B., Duan, P., Xu, Y., Wang, F., & Fan, Y. (2013). Co-liquefaction of micro-and macroalgae in subcritical water. *Bioresource technology, 149*, 103–110.
14. Kong, W., Shen, B., Lyu, H., Kong, J., Ma, J., Wang, Z., & Feng, S. (2021). Review on carbon dioxide fixation coupled with nutrients removal from wastewater by microalgae. *Journal of Cleaner Production, 292*, 125975. https://doi.org/10.1016/j.jclepro.2021.125975.
15. Menetrez, M. Y. (2012). An overview of algae biofuel production and potential environmental impact. *Environmental science & technology, 46*(13), 7073–7085.
16. Mostafa, S. S. (2012). Microalgal biotechnology: prospects and applications. *Plant science, 12*, 276–314.

17. Rasoul-Amini, S., Ghasemi, Y., Morowvat, M. H., Ghoshoon, M. B., Raee, M. J., Mosavi-Azam, S. B., ... & Khoubani, S. (2010). Characterization of hydrocortisone bioconversion and 16S RNA gene in Synechococcus nidulans cultures. *Applied Biochemistry and Microbiology*, *46*, 191–197.
18. The Intergovernmental Panel on Climate Change (IPCC) (2011). Special Report on Renewable Energy Sources and Climate Change Mitigation. Intergovernmental Panel on Climate Change. https://www.ipcc.ch/report/renewable-energy-sources-and-climate-change-mitigation/.
19. Tirichine, L., & Bowler, C. (2011). Decoding algal genomes: tracing back the history of photosynthetic life on Earth. *The Plant Journal*, *66*(1), 45–57.
20. Williams, C., and P. Porter. (2011). Introduction to Bioenergy. Module 1 in S. Lezberg and J. Mullins (eds.) Bioenergy and Sustainability Course. On-line curriculum. Bioenergy Training Center. http://fyi.uwex.edu/biotrainingcenter.
21. Yap, J. K., Sankaran, R., Chew, K. W., Munawaroh, H. S. H., Ho, S. H., Banu, J. R., & Show, P. L. (2021). Advancement of green technologies: A comprehensive review on the potential application of microalgae biomass. *Chemosphere*, *281*, 130886. https://doi.org/10.1016/j.chemosphere.2021.130886.
22. Zhang, R., Parniakov, O., Grimi, N., Lebovka, N., Marchal, L., & Vorobiev, E. (2019). Emerging techniques for cell disruption and extraction of valuable bio-molecules of microalgae Nannochloropsis sp. *Bioprocess and biosystems engineering*, *42*, 173–186. https://doi.org/10.1007/s00449-018-2038-5.

Microalgal Biomass Production and Harvesting

2

Abstract

Microalgae are unicellular photosynthetic organisms that utilize solar energy through photosynthesis turn into chemical energy (Gupta and Bux in Springer Nature Switzerland AG, Domestic and industrial wastewater treatment, 2019 [13]). They are emerging technologies for producing high-value metabolites like vitamins, carbohydrates, essential fatty acids, lipids, proteins, and minerals for diet and nutraceutical industries. They have higher photosynthesis efficiency, higher growth rate, and can be grown in barren lands with minimum contamination risks. Microalgae can also be used for wastewater treatment, phytoremediation and CO_2 sequestration, with numerous environmental uses (Udayan et al. in Bioresource technology 344, 2022 [23]). Efficient microalgal culture methods proficient of producing fairly higher microalgal biomass with high photosynthetic efficiency are attracting increased study consideration, necessitating further technical progress. Microalgal cultivation with industrial waste water has been demonstrated to be an efficient cost reduction approach. The most of research on mixotrophic and heterotrophic culture has been conducted in laboratories. Microalgal culture and harvesting with high throughput have led to an dynamic field of study in recent. However, in spite of significant breakthroughs in research, further studies are required to enhance higher biomass output for industrial purposes. It is also vital to optimize the microalgal harvesting process, as it now showing around 30% of overall production costs. Currently, there is more research on enhancing lipids and other high-value metabolite synthesis from microalgae instead of biomass collection and water removal (Udayan et al. in Bioresource technology 344, 2022 [23]). An effective and optimum harvesting method must be adaptable to varied microalgae and achieve high biomass retrieval with the low energy and running costs. Different harvesting strategies have been utilized, including biological, chemical, physical and electrical harvesting, and in certain situations, two or more approaches are collectively used to maximize

H. N. Akolkar et al., *Bioenergy from Microalgal Biomass*, Synthesis Lectures on Renewable Energy Technologies, https://doi.org/10.1007/978-3-032-03434-2_2

biomass retrieval from the growth media. Several research on biomass collection have lately become accessible. However, fundamental comprehension and information about harvesting procedures remain unclear. However, the purpose of this research was to examine the several harvesting methods as well as its advantages and disadvantages from an industrial viewpoint.

Keywords

Microalgae • Biomass production • Harvesting • Photobioreactors • Cultivation conditions • Strain selection • Biofuels

2.1 Microalgal Strains and Cultivation Methods

Microalgae are key raw-material for modern generation biofuels, which thrive in freshwater, wastewater, seawater, food waste, and also on saline-alkaline soil, producing more biomass than land plants while requiring less resources. Microalgae grow quickly, have a greater photosynthetic efficiency, and need less area, making them suitable for producing a variety of biofuels. Over the last two decades, researchers have achieved considerable advances in the study of microalgal biodiesel. Numerous species, including *Chlorella vulgaris, Botryococcus, Desmodesmus, Dunaliella salina, Nannochloropsis oceanica, Neochloris, Tetraselmis* and *Scenedesmus* have been tested as perfect choice for biodiesel production, with some produces biodiesel on a laboratory scale [27].

Synechococcus spp. also meet EN 14214 criteria for cetane number and iodine value. Furthermore, the fuel qualities of a diesel and microalgal biodiesel blend are widely investigated [27]. While several research have established techniques for obtaining lipid from microalgae by varying growth conditions, basic limits cannot be dazed if inappropriate strains are selected for production of biofuel. It is critical to do extensive research on species-specific aspects of lipid synthesis in microalgae. As a result, selecting robust strains capable of producing a large volume of desired products is an important stage in large-scale biofuel generation.

Microalgae species vary in size, amount of lipid and fatty acid (FA) present, hence not each one is suitable for the biofuel generation. Researchers have presently shown an interest in genetically edited microalgae, and various species are changed, such as *B. braunii, C. vulgaris, Chlamydomonas, D. salina, Galdieria sulphuraria, Phaeodactylum tricornutum, Thalassiosira pseudonana,* and others. This method uses genetic engineering for improvement and modify the standard features of microalgae, producing of high-amount of lipids. It outperforms random mutagenesis in terms of effectiveness, precision, and speed, and has several potential applications.

While the commercial ability of microalgal biofuels is generally acknowledged, various biotechnological, economic and environmental difficulties must be overcome before microalgae harvests may be introduced into the worldwide market. These comprise cost-effective harvesting and dehydration technologies, process advancements in harvest extraction, and a detailed investigation of genetic tools along with genomic resources (Table 2.1).

2.2 Different Microalgal Cultivation Methods

There are various cultivation methods for microalgal production [4, 7, 17, 20, 24]. Microalgae cultivation looks to be quite easy, with little or no oversight, utilizing basic fertilizers and even waste water to absorb CO_2 from the environment. Large-scale microalgal farming can take place in whichever open or closed systems. Open systems include variety of settings, such as tiny artificial ponds and enormous amounts of exposed water, all of which use different agitation mechanisms [16]. Closed systems encompass a wide range of PBRs, such as plastic bags, vertical or horizontal tubes, airlift reactors, bubble columns, flat panel tubular reactors, membrane-based systems and plate reactors. At first, open systems considered commercially viable than PBRs because they were less expensive to install, could handle larger regions, and were simple to manage.

Yet, these systems are impotent to manage constant ambient light and temperature parameters and are naturally prone to microbial and fungal infection, resulting in poor light utilization by cells, predator contamination, evaporation loss, and CO_2 escape. Because of the drawbacks of open systems, closed PBRs have received a lot of attention. On the one hand, PBRs inhibit evaporation, which reduces contamination concerns and CO_2 losses. On the other side, PBRs have great efficiency of photosynthesis, resulting in higher biomass and lipid output.

2.2.1 Photobioreactors (PBRS)

PBRs can grow a broader range of microalgal species, whereas open systems generally cultivate only a fewer stronger photoautotrophic microalga. *Chlorella pyrenoidosa, C. reinhardtii, Golenkinia sp., N. oceanica,* and other microalgae have been grown effectively in PBRs. Few of them demonstrate better growth performance inside PBRs than in open systems, demonstrating the potential to accumulate more lipids and biomass [27]. Because the design of different PBRs has a direct impact on biomass production, there is a growing interest in developing new PBRs to enhance process design and operation. By redesigning PBR setups, researchers were able to maximize biomass output, enhance pollutants removal efficiency, minimize costs, and increase space domination. Among these, investigations on improving lipid synthesis and quality using membrane PBRs mostly attempt to employ wastewater to induce lipid production. This technique to water resource management is thought to be more ecologically benign than other forms of PBRs.

Table 2.1 Different microalgal strains [27]

Species	Biomass (g L^{-1})	Productivity of biomass (mg L^{-1} d^{-1})	Content of lipid (%, W/WDW)	Productivity of lipid (mg L^{-1} d^{-1})
Ankistrodesmus braunii	1.43–1.57	–	22.1–34.4	–
Ankistrodesmus falcatus	1.4–2.29	340	16.49–53	56.07
Ankistrodesmus fusiformis	0.93–3.04	62.3–240	20.66–38.28	18.94–116.88
Botryococcus braunii	0.5–2.74	104–250	20.4–74.4	48.3–112.43
Chlamydomonas sp.	2.49–6.09	240–280	4–46.2	36.17–380
Chromochloris zofingiensis (Chlorella zofingiensis)	8.3	171.4–691.67	30–64.5	36–445.7
Chlorella protothecoides	–	–	11.1–42.3	15.25
Chlorella sorokiniana	–	43–100	44	22.2
C. vulgaris	2.83–3.47	234–730	11.9–54.2	30.2–204.91
Chlorococcum sp.	–	–	20.77–38.9	22
Desmodesmus brasiliensis	–	130	17.99–24.2	23.39
D. salina	1.23	50–150	18.9–25.1	10.26–36.48
Dunaliella sp.	–	150	22–33	33–39.3
Dunaliella tertiolecta	1.3	81	18–50	8.4–20.2
Haematococcus pluvialis	0.57–0.79	80.71–112.14	13–35.9	–
Nannochloris sp.	0.17–0.75	18.1–83.44	26.29–53.13	29.21–59.03
Neochloris oleoabundans (Ettlia oleoabundans)	–	386.5–413.1	13.1–74.5	46.7–172.6
Raphidocelis subcapitata (Pseudokirchneriella subcapitata)	–	80	16.3–28.43	22.74
Scenedesmus dimorphus	0.75–1.16	44–68	15.14–29.15	15.449

(continued)

Table 2.1 (continued)

Species	Biomass (g L^{-1})	Productivity of biomass (mg L^{-1} d^{-1})	Content of lipid (%, W/WDW)	Productivity of lipid (mg L^{-1} d^{-1})
Scenedesmus incrassatulus	0.99–3.95	–	19.7–37.7	–
Tetradesmus obliquus (Scenedesmus obliquus)	0.5–1.5	50–160	16.73–55	26.77

PBRs enhance the efficiency more than closed systems, with which microalgal cells utilize light energy due to its exceptional light path architecture. The total photosensitive conditions in PBRs are assessed by measuring penetration and light dispersion. Here, improving the ratio of the baffle aperture gap to the bioreactor width may considerably improve microalgae's mass transfer and light intake efficiency in flat-plate PBRs.

Microalgae effectively transform optical light into chemical energy, resulting in the resourceful accumulation of lipids and biomass. Actually, investigations have shown that microalga grown in upgraded PBRs have approximately tripled biomass and lipid accumulation. Furthermore, hydrodynamics has a significant impact on PBR performance. Mixing of proper fluids help to avoid sedimentation, controls nutrition supply, and also directs the microalgal cells mobility in the reactor, all of which have an impact on the exposure of light. When, four diverse PBRs were tested at variable air flow rates to determine its influence on *N. oleoabundans* algal lipid and biomass production, and it was discovered a ring sparger having bubble column reactor with performed best.

By incorporating mixers and enhancing sparger performance, innovative PBRs with high gas–liquid mass transfer rates, efficient mixing and well-defined patterned fluid flow may be created for microalgae lipid synthesis and biofuel generation [27]. The findings imply that reactor configuration and operational factors impact microalgal biofuel generation in PBRs. Ideally, PBRs should not disrupt or destroy cells; yet, turbulent flow damages cell in PBRs. Thus, enhancing the performance of PBRs is a critical technological technique for increasing microalgae biomass and lipid production.

Types:

- **Tubular PBRs**: They consist of transparent tubes arranged horizontally or vertically.
- **Flat Plate PBRs**: This technique utilizes flat panels to maximize light exposure.
- **Column PBRs**: In this method, vertical columns that allow efficient gas exchange and mixing.

2.2.2 Open Pond Systems

The open pond is the utmost basic technique for microalgal culture and mass production, in which microalga grow in settings just like their natural habitat [21]. Circular ponds race-way ponds and unstirred ponds are common commercial-scale open systems. Open ponds provide the benefit of providing CO_2 to microalgae due to the large contact between air and liquid culture. It offers additional advantages such as ease of scaling up, less energy utilization, and inexpensive construction and running expenses; so, open ponds are frequently utilized in microalgae mass culture.

Mixing and CO_2 transfer are two factors that influence microalgae development in open ponds; paddle wheels are commonly used to control nutrient content, temperature and CO_2 supply through the open pond by increasing mass allocation (Fig. 2.1).

Diverse microorganisms cause biological contamination that cohabit with cultivated microalgae and strive for nutrient provided. Production in an open pond is restricted to microalgae which can withstand harsh ecological conditions since the total culture parameters are hard enough to maintain as per the seasonal variation. One significant drawback of open ponds is the increased danger of air pollution. The primary pollution in open pond algal cultivation is biotic contamination generated by a variety of bacteria. The large boundary in between open pond with liquid culture and air promotes CO_2 intake, releases excess dissolved O_2, and offers a suitable home for a variety of biological pollutants [27].

Other biological pollutants include microalgae, zooplankton, fungus, bacteria, and viruses. These biotic pollutants can reduce biomass quality or yield, lowering the efficiency of desirable harvests such astaxanthin; so, measures for detecting and removing biotic contaminants during the cultivation are necessary.

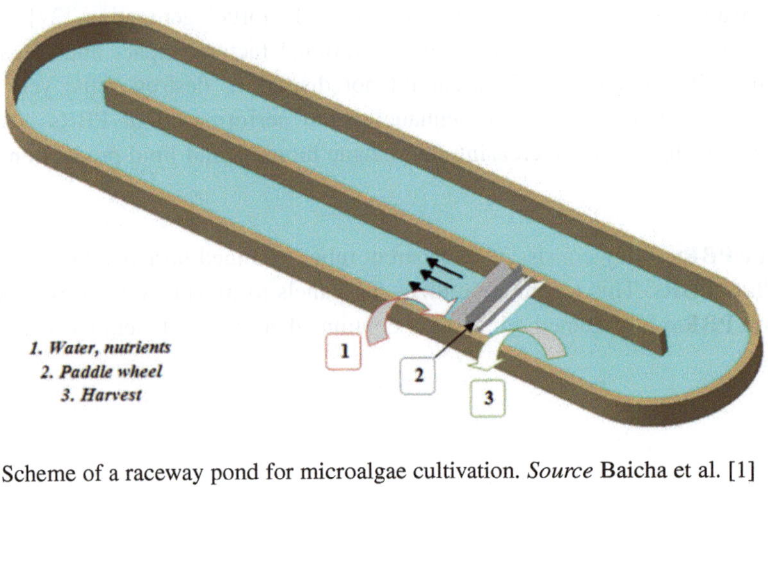

Fig. 2.1 Scheme of a raceway pond for microalgae cultivation. *Source* Baicha et al. [1]

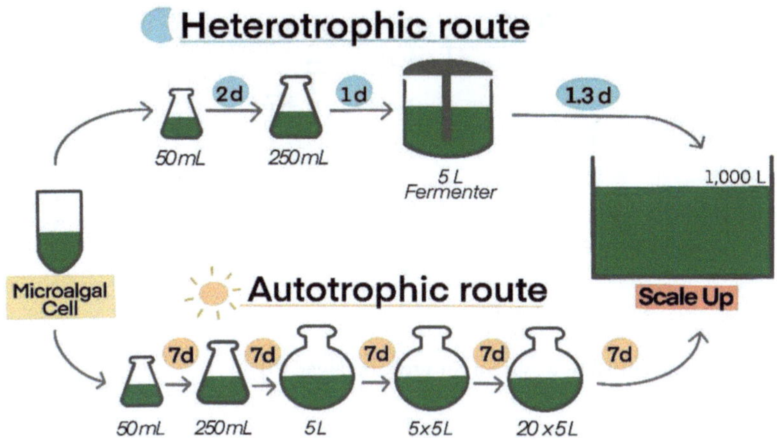

Fig. 2.2 Time consumption comparison between heterotrophic and autotrophic routes of *C. vulgaris* [2]

2.2.3 Heterotrophic Cultivation

Algae are classified as autotrophic organisms, meaning they rely on solar radiation for energy to thrive. However, several strains of microscopic algae do not require light energy if they have another readily available source of energy, such as sugars. Heterotrophic growth refers to the development of microalgae without light using simple organic materials as energy. Furthermore, heterotrophic culture is metabolically favorable for increased lipid buildup, which would be advantageous in biodiesel generation.

Heterotrophic culture provides numerous significant benefits that need careful cultivation and continual monitoring of sterile conditions.

- Cultivation is independent of weather, climate, or light sources.
- In temperate regions, it may be used all year.
- Microalgae are grown under regulated circumstances to prevent contamination.
- High crop density can minimize harvesting expenses.

A spray drier, algal cell disintegration, and different separation centrifuges are also required for heterotrophic cultivation (Fig. 2.2).

2.2.4 Mixotrophic Cultivation

Mixotrophy varies by species of microalgae. In rare circumstances, mixotrophic microalgal culture can be more productive than autotrophic and heterotrophic cultivation

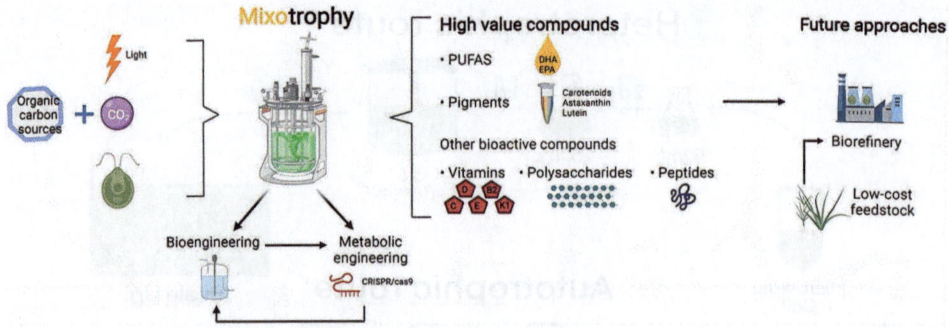

Fig. 2.3 Mixotrophic cultivation system for growing algae [5]

separately, although it is usually within these two ranges. Many microalgae species are mixotrophic, meaning they can both photosynthesis and consume organic materials. Mixotrophic microalgae have the potential to employ either autotrophic or heterotrophic metabolism, allowing for cell development independent of sunlight. There is no necessity of light for mixotrophic microalgal culture, therefore algal biomass may be generated with either organic carbon or light as the source of energy.

Mixotrophic cultivation process is controlled by the organic substrates such as glucose throughout both the dark and light phases, resulting in decreased biomass decrease through the dark phase. Mixotrophic microalgae have higher growth rates than autotrophic and heterotrophic cultivations because they can integrate both heterotrophic and autotrophic mechanisms during the 24-h set (Fig. 2.3).

2.2.5 High-Rate Algal Ponds (HRAPs)

The HRAPs are one of the modified stabilizing ponds which promotes algal proliferation to create an atmosphere conducive to growth of microbes and reproduction, which results in a strong bacterial-algal symbiosis. HRAPs are planned to be race-track-style or open-channel ponds. A single paddle wheel gently circulates wastewater around the HRAP at rates ranging from 0.15 to 0.3 m per second. The paddle wheels in the water column generate turbulent eddies, which cause vertical mixing. Microalgae may grow to high biomass concentrations because to the shallow water depth, which ranges between 0.2 and 0.7 m, along with vertical mixup [14].

When compared with standard biological treatment methods, this innovation offers the benefits of minimal operating expenses and investment; it does not require huge expanses of land to be employed, and the procedure is straightforward. HRAPs outperform traditional ponds in terms of nutrient removal and natural disinfection, as well as recovery of resource in the form of microalgal biomass for valuable reuse as biofuels, feed or fertilizer. Furthermore, HRAPs have significantly higher dissolved oxygen levels compared to standard stable ponds. The key variables that influence DO in HRAPs are the alternation of respiration and photosynthesis of algae, atmospheric dissolution and bacterial respiration.

HRAPs remove pollutants largely through synergy in physiological activities between algae and bacteria. Algae utilize solar light as an energy source, a carbon source i.e., CO_2, and a nitrogen source—ammonia to reproduce algal cells through photosynthesis using chlorophyll [15]. They also release increased amounts of Oxygen, which bacteria use to degrade organic matter, resulting in low energy and carbon utilization. Aerobic bacteria breakdown wastewater organic materials through oxidation, turning them in tiny inorganic compounds and creating CO_2. HRAP systems may be utilized to build a sustainable wastewater treatment, making them a popular wastewater treatment innovation having a lot of potential for growth.

2.2.6 Attached Growth Systems

The connected culture approach has been researched in recent years with the goal of lowering the cost of microalgae harvest. Attached culture allowed microalgal cells to concentrate on substratum surfaces, making microalgae-water separation simpler. Existing connected culture methods are classified into three types based on the position of the microalgal cells and culturing media on the surface of substratum: permanently submerged, intermittently submerged, and perfused [10, 26]. In many systems, microalgae likely to be collected by scratching the surface of substratum. Attached cultivation methods, as opposed to suspended cultivation systems, made biomass gathering simpler.

Furthermore, solid microalgal biomass composition in typical attached growing methods can reach up to10-20%, which is preferable to traditional suspended microalgal cultivation, where the biomass content of 0.5–5 g/L equates to a solid concentration of just 0.5% or less. Attached growth methods are particularly advantageous since they have a higher light use efficiency than suspended culture, as established in earlier research. Other advantages include a large reduction in footprint size, excellent Carbon Dioxide mass transfer efficiency, the separation of solid retention time (SRT) and hydraulic retention time (HRT) and so on.

Despite all of the benefits outlined overhead, an attached cultivation system with both a simple design and a low cost is not yet existing. To address this issue, this study presented a unique connected cultivation method, known as a capillary-driven photobioreactor (CPBR), in which capillary action is used to carry the culture media, saving energy. Concentrated attached biomass may be collected from CPBR by squeezing/pressing the carrier, making it easy to recycle.

2.3 Biomass Harvesting Techniques

The algal group has different properties and includes both bacterial cyanobacteria and eukaryotic algae. Microalgae species have been used in a variety of applications, particularly human and animal nutrition, bioremediation, medicine, and biofertilizers. Number of studies have recently demonstrated usefulness of microalgae biomass in biofuel generation. Furthermore, because microalgae show a high CO_2 fixation efficiency throughout development, barren land is essential for their cultivation. The process of microalgae biomass producing consists of three steps: growing, harvesting, and processing. The harvesting process is regarded tough because of its huge cost, and haning a direct impact on the processing stage [11]. Furthermore, various elements impact the harvesting process, such as microalgae cell size, cultural circumstances, the electronegative characteristic of the growth rate, cell membrane, and so on. Harvesting microalgal biomass is a complex procedure that requires a variety of chemical and mechanical methods. A variety of harvesting methologies have been used to extract biomass, including chemical and bio-flocculation, membrane filtration, coagulation flotation and centrifugation [9].

2.3.1 Filtration

Filtration is a significant physical procedure utilized for separation of particles from liquids, allowing only fluid to pass. To extract microalgae biomass, many filtration procedures are employed, such as microfiltration, ultrafiltration, pressure filtration, vacuum filtration and so on. Membranes are important in the filtration technique for harvesting microalgae because membrane fouling is a major concern with expensive harvesting procedures associated with low fluxes. To reduce fouling development, a surface-coating substance such as hydrophilic polyvinyl alcohol polymer can be used. The polyvinyl alcohol coating increased hydrophilicity and performance of the membrane surface by increasing maximum flow—36%, with a out of recovery rate. Microalgae derived transparent exopolymer particles are employed to minimize membrane fouling in various filtering processes [9].

To collect microalgal cells in the crossflow membrane filtering method, a microfiltration membrane (0.2 μm polyvinylidene fluoride) and three ultrafiltration membranes viz. regenerated cellulose, hydrophilic polyethersulfone and polyethersulfone were utilized [9]. Hydrophilic polyethersulfone—UH050 membranes performed better as membrane fouling, crossflow velocity and transmembrane pressure.

To filter microalgal biomass, many types of permeable membranes were utilized during the membrane filtration process. Previously, various writers studied the influence of ultrafiltration and microfiltration membranes on microalgae biomass collection. Membrane filtering processes can be enhanced using varying pore sizes. Serious concern in membrane filtration is fouling, because pores clog.

Thus, pore size is the primary parameter used to classify ultrafiltration (2–100 nm), microfiltration (100–10,000 nm), and nanofiltration (0.5–2 nm) membranes [28]. Generally, microfiltration membranes have an extensive variety of pore sizes, whereas ultrafiltration membrane shows a narrow pore size [9]. Nanofiltration membrane have the smallest pore size. Furthermore, if membrane is with smaller pore sizes possess slower rate of filtration. Polyacrylonitrile, polyethersulfone, polyvinylidene fluoride, polyvinyl chloride and polytetrafluoroethylene are widely used membranes.

2.3.2 Flocculation

Flocculation is a frequently used method for harvesting microalgal biomass. This technique collects dispersed cells present in media and settles them down using different chemicals and bio-flocculants.

2.3.2.1 Chemical Flocculation

Chemical flocculation is recognized as an effective method for collecting microalgae. The Mg-sericite flocculant was also used to collect *C. vulgaris*. The pH of the growing medium has a substantial role in harvesting efficiency. To collect *C. vulgaris*, combine flocculants with calcium oxide and poly γ-glutamic acid. The results showed that employing mixed flocculants greatly reduced flocculant concentrations. Polydiallyldimethylammonium chloride is the most efficient flocculant when compared to chitosan and Superfloc®, at a dose of 5 mg/L, achieving 90% flocculation efficiency at pH10 per hour. Furthermore, above flocculant significantly improved *N. salina* harvest by filtering. Also use of brucite or $Mg(OH)_2$ to collect *Phaeodactylum tricornutum*, a diatom have been reported [9].

In a continuous repeated media, it was discovered that $FeCl_3$ used flocculation with had a considerable influence on biomass of *C. vulgaris* without changing its lipid content. Similarly, *Scenedesmus* sp. was isolated from BG-11 medium and ferric chloride containing wastewater. Harvesting efficiency enhanced when the culture pH was reduced below 6.5. One study discovered that flocculation which is microwave-assisted significantly lowered the content of culture media flocculants [9].

A sulfuric acid and metal coagulant $Fe_2(SO_4)_3$ were also used to extract *Chlorella* sp. KR-1. Furthermore, *Chlorella* sp. KR-1 biomass treated with the acid might be used directly for lipid extraction without the inclusion of a catalyst [18]. At pH 9, harvesting *D. salina* using the flocculant $FeCl_3{\cdot}6H_2O$ (0.50 mM) proved efficient. *Arthrospira* maximum biomass was efficiently extracted by flocculation with $CaCl_2$ concentrations extending from 0.2 to 2.0 g/L. *Chlorella* microalgal biomass is harvested using a handmade helical tube. In this procedure, a recovery over 90% was attained utilizing a 0.12 g/L diluted algal solution with sedimentation and flocculation. To harvest microalgal biomass, the addition of $CaCO_3$ in the culture media is another successful method.

C. vulgaris biomass may be harvested using the alum flocculation process. Another study found that the highest flocculation effectiveness more than 90%, was achieved with a flocculant concentration of 0.5 g/L and a growth medium pH of 8.2. Ferric chloride and aluminum chloride—chloride flocculants and ferric sulfate and aluminum sulfate—sulfate flocculants used to harvest *C. vulgaris.* The findings showed that the flocculation effectiveness of above flocculants ranged between 93.5 and 98.8% at concentrations of 100 mg of chloride salts and 60 mg of sulfate salts per liter of microalgal culture. Furthermore, the flocculants utilized did not change the biomass contents.

In one flocculation procedure, alkaline pH influenced the shape of collected cells of *C. vulgaris, Phaeodactylum tricornutum,* and *Porphyridium purpureum.* Furthermore, the inclusion of polyacrylamide resulted in greater harvesting efficiency and preservation of industrially essential chemical components.

2.3.2.2 Electro-Flocculation

In one research, several flocculation methods, including chemical flocculation, auto-flocculation and electrolytic flocculation, used to harvest *Synechocystis* PCC 6803 and *Chlorella* sp. MJ 11/11. From these, electro-flocculation had the best flocculation effectiveness at 98%. One more research used aluminum electrolysis and electro-flocculation to collect *C. vulgaris.* When the current density was increased, microalgae were collected more quickly. *D. Salina* harvesting was studied using a successful electro-flocculation technique combined with local sand. In comparison to conventional electro-flocculation procedures, our technology reduced average electrical energy usage by 51.03%. In the electro-flocculation procedure, an battery consisting aluminum-air was effectively used to collect *D. salina.* Microalgae are harvested using an electrolytic flotation technique without the use of flocculants. In this procedure, carbon—anode and stainless steel—cathode, were employed based on harvesting efficiency. The flocculation which is alkali-induced combined with an electrolysis called as salt bridge electro-flocculation technique was successful for harvesting microalgal biomass. Furthermore, the salt bridge substantially reduced microalgae cell damage caused by anode oxidation, and the algal biomass was free of external contaminants. The aluminum electrodes demonstrated better efficiency of harvesting and the method had no influence on the levels of *C. vulgaris* pigments and lipids [9].

2.3.2.3 Bio-Flocculation

Among the many flocculation options, bio-flocculation is an appealing approach for harvesting microalgae while avoiding negative effects induced by chemicals. Cationic polyelectrolyte—chitosan is widely utilized to harvest numerous algae species due to its biodegradable and non-toxic qualities. However, chitosan cost is prohibitively expensive, prompting research into other bio-flocculants. The addition of walnut protein extract boosted the flocculation effectiveness of chitosan (6 mg/L) by up to 98%. A comparison analysis found that the optimum chitosan concentration of 0.25 g/L yielded more than

90% of microalgal biomass [12]. *C. sorokiniana* was collected employing chitosan, with flocculation efficacy exceeding 99% at pH levels below 7.

Greenfloc 120, Cationic starch also utilized to collect *C. protothecoides*, with the highest efficiency of flocculation achieved at pH 7.7–10. *C. sorokiniana* cultures were effectively flocculated with cationic nanocrystalline cellulose at concentrations both above and below the isoelectric threshold. A cationic biopolymer—α-Poly-l-lysine was used to collect microalgae. The bio-flocculant's natural antimicrobial action efficiently prevented biological contamination. Chitosan had a better flocculation efficiency than polyacrylamide. Furthermore, no alterations were seen in the chemical makeup of microalgal biomass.

The bio-flocculation effectiveness of leftover brewer's yeast was investigated to cultivate *C. vulgaris*. 2-chloro-N,N-diethylethylamine hydrochloride was employed to chemically alter yeast followed after hydrolysis, and the flocculation output for harvesting *C. vulgaris* was measured. Poly-γ-glutamic acid was used to harvest *C. protothecoides* and *C. vulgaris*. Furthermore, the study showed the extracted microalgal cells were not damaged, resulting in no lipid costed throughout the flocculation method. A study found that using poly-γ-glutamic acid from *Bacillus licheniformis* helped harvest *Desmodesmus brasiliensis*. Another research discovered that *Trichoderma reesei* QM 9414, the filamentous fungus has a high pellet-forming capacity for collecting species of *Scenedesmus* [9].

Co-cultivation of *C. vulgaris* and *Aspergillus niger* with 2 g/L glucose resulted in a harvesting efficiency of more than 90%. The findings revealed that the carbon supply was critical for improving fungal growth and producing cell pellets. The combined cultivation of *C. sorokiniana* and *Aspergillus flavus* was shown to be the great effective in forming pellets in BG11 with nutrient-supplement. *Aspergillus niger* was also used as a bio-flocculant when harvesting microalgal biomass. The bio-flocculant shown capacity to respond to a wide range of pH (3.0–9.0). The metabolites generated in the media may be accountable for its bio-flocculation feature. *Chlorella sp.* was also harvested using fungal spores and pellets. In this technique, co-cultivating *Penicillium sp.* spores or pellets with *Chlorella sp.* resulted in the best flocculation efficiency (99%).

The efficiency of flocculation of cellulose nanocrystal cations for cultivation of *N. oculata* was studied. The natural organic flocculant tannin S5T demonstrated the similar flocculation effectiveness in the examined microalga. Another research used four natural flocculants—chitosan, *Moringa oleifera,* cationic starch and Tanfloc—to harvest *S. obliquus* and *C. vulgaris*. Tanfloc had the highest harvesting effectiveness (98%) for *S. obliquus* at 20 mg/L and *C. vulgaris* at 30 mg/L.

Seed powder of *Strychnos potatorum* also utilized to cutivate *C. vulgaris*, with the maximum efficiency (99.68%) achieved with established conditions of bioflocculant concentration 100 mg/L, 35 °C, 150 rpm, and a 30-min incubation period. Cationic starch outperformed other chemical coagulants in terms of microalgae biomass output throughout a wide pH range. In an acidic pH range, seed powders of *Hibiscus esculentus* and

Moringa oleifera increased biomass productivity by up to 50%. Polymeric extracellular compounds derived from *S. acuminatus* were employed to collect the similr algae using the flocculation technique. The study showed that adding flocculant, an extracellular polymeric at 3.2 mg/g significantly reduced the need of coagulant—alum (Al^{3+}) from 4.5 to 77.6 mg/g [9].

2.3.3 Flotation Centrifugation

Flotation has recently gained popularity as a method of removing microalgae from the suspension. Flotation method converts gas or air into bubbles using a liquid/solid solution. As a result, solid contents in the media become bonded to gaseous molecules, accumulating to surface. The flotation process is classified into three forms based on the bubble size: dissolved and dispersed air, electrolytic, and ozone diffused flotation techniques.

Flotation with dispersed air along chitosan and saponin was an excellent method for harvesting *Scenedesmus obliquus* and *C. vulgaris.* Hexadecyltrimethyl-ammonium bromide, a surfactant, outperformed tea saponin in terms of *C. vulgaris* flotation harvesting efficiency. In this approach, hexadecyltrimethylammonium bromide defused the potential of algae, whilst tea saponin converted the hydrophilic microalgal surface to hydrophobic. The study showed that tea saponin surfactants and hexadecyltrimethylammonium bromide might improve the affinity between bubbles and *C. vulgaris*, letting collection of the microalgae more readily [18]. In case of *C. sorokiniana*, pH manipulation via the flotation technique with dissolved air is regarded as an effective way for wastewater microalgae biomass harvesting. The dissolved air flotation method with biopolymer flocculant (AFlok-BP1) based on tannin for marine *Nannochloropsis* sp. showed a better harvesting efficiency.

Buoy-bead flotation procedure is commonly used to enhance microalgal harvesting efficiency and decrease the chemical use. Using this technology, polymeric microspheres with surface-layer created to collect *C. vulgaris.* Findings showed that best harvesting efficiency (98.43%) was achieved at concentration of 0.7 g/L of surface-layered polymeric microspheres which may be successfully recycled for harvesting. Another research found that the buoy-bead flotation approach had an 89.9% harvesting rate. Furthermore, the scientists discovered that pH, microsphere diameter, and agitation speed had a significant effect on harvesting efficiency of microalgae. Another report shows that sodium-alginate microspheres were used to collect *C. vulgaris.* The findings revealed that sodium-alginate microspheres mixed with little amount of aluminum sulfate had a greater harvesting efficiency than buoy-bead flotation and air flotation approaches.

The thermal flotation technique was employed to collect *S. obliquus* and *C. vulgaris.* The study revealed that *S. obliquus* had a greater harvesting efficiency (88.16%) than *C. vulgaris* (47.16%) since the hot pre-flocculation procedure destroyed the lipids, proteins and carbohydrates on cell surfaces of microalgae. *Dunaliella salina* is an important microalgal species utilized in industry due to its capacity to produce more carotenoid pigments. To harvest this species, the flotation procedure with NaOH was utilized.

Recently, colloidal gas aphron technology is being applied in flotation. This stable colloidal gas aphrons approach was utilized to eliminate *N. oculata*, *Arthrospira platensis* and *C. vulgaris*. To prevent chemical residues during microalgae harvesting, a bio-flotation approach was used that used oil emulsion and *Moringa* protein extract. This approach achieved 86.5% harvesting efficiency for *Nannochloropsis* sp. In comparison with monomeric cetyltrimethylammonium bromide, the new gemini surfactant outperformed the former. To harvest *Chlorella* sp., Al^{3+} coupled along cetyltrimethylammonium bromide increased algal floc size and hydrophobicity, aiding flotation.

Flotation is then paired with centrifugation, which separates particles of differing densities or sizes using centrifugal force. Centrifugation is used to distinguish microalgal cells from the culture media during microalgal biomass harvesting [25]. Centrifugation is based upon density difference between microalgal cells and the culture media [22]. When centrifugal force is applied, the microalgal biomass is forced to the base of the centrifuge tube, leaving the culture material on top.

2.3.4 Electrocoagulation

Electrocoagulation is one of the greatest modern methods for extracting microalgae biomass. The treatment with graphite electrode (67.44%) had the lower harvesting efficiency for *N. oculata*. In contrast, utilizing aluminum and iron electrodes resulted in a better harvesting efficiency of *N. oculata*. Different electrodes have a substantial effect on the biochemical composition of *N. oculata* during electrocoagulation. To minimize metallic electrode depletion, non-sacrificial electrodes are used to collect *S. obliquus*. Stainless steel -cathode and carbon fabric-anode were employed to collect microalgae instead of aluminum electrodes, and which demonstrated the best flotation efficiency (98%) and due to their less corrosive qualities shows lowest pollution.

A novel electrocoagulation electrode using dielectrophoretic force was tried for harvesting *Tetraselmis* sp. In electrocoagulation, *Tetraselmis* sp. was harvested using a novel cylindrical interdigitated electrode array. Furthermore, the microalgae biomass harvesting efficiency rose by 96.18% while the distance between electrode was reduced to 0.5 cm. Continuous electrocoagulation method was established to collect *C. vulgaris*. By using this technique more than 85% harvesting efficiency is achieved with a minimum degree of Iron contamination. The microalgae harvesting efficiency using electroflotation without any coagulation is also investigated in terms of various phases of hydrophobicity. The hydrophobicity of microalgae was key in the electroflotation process. The harvesting efficiency of *Pandorina sp.* (10%) and *Scenedesmus sp.* (70.1%) was significantly lower. Among the examined microalgae, *Tribonema* sp. had a harvesting efficiency of up to 96.2%.

2.4 Biomass Pre-treatment Methods

The wide range of microalgae and synthesized products necessitates a various pretreatments; there is no proper method that is utilized on all microalgal biomasses [8]. However, most strategies may be divided into four groups based on the required product:

(1) Drying, (2) thermochemical processing, (3) cell disruption and (4) targeted fractionation.

2.4.1 Drying

Drying is used for pretreatment for subsequent extraction or simply to stabilize algal biomass since the reduced activity of water minimizes microbiological and chemical degradation. The process comprises mostly removal of water, which is dependent on transmission of heat and water dispersion. Other factors, including as denaturation and mechanical impacts, can have an effect on biomass components and the cell structure [19]. This can have a variety of implications in post-processing, including improved digestibility, protein precipitation, cell collapse and enzyme deactivation. When the dried algal biomass is harvested for useful compounds just like enzymes or vitamins, a mild and gradual drying method is required because it generates porous viable, algal biomass with minimal modification of molecules. Lyophilization might be a useful option. When biomass is treated to extract more resistant remains, the drying rate and temperature can be increased. Recent study indicates that traditional techniques such as spray drying might increase permeability of cell, presumably due to an enlargement effect in the fast drying.

Several factors occur throughout drying that might be investigated to alter the microalgal biomass structure. The biomass is heated by contact with a hot air—convection, solid surface—conduction or electromagnetic radiation such as infrared or microwave. Heat is required for evaporation of water, which is subsequently evacuated by an air current. A vacuum can be employed to improve mass allocation, but there is requirement of energy for phase shift, so the frozen biomass must be heated during lyophilization. The technique for drying microalgal biomass is often chosen based on intended application and production scale. Energy costs in the drying technique surpass those in microalgae biomass production, particularly when the water is decreased to around 10%, as during its usage for oil extraction. It can cause up to 75% of the total algal processing cost [6].

Spray-drying and freeze-drying are two of the costliest drying processes, yet when microstructure like features of dried microalgal biomass are examined, they are frequently shown to be superior than traditional techniques such as drum drying. *Chlorella* and *Spirulina* spray-dried particles have been shaped like spheres with a centered void. Each particle of *Chlorella* was made up of thousand cells, while *Spirulina* had numerous trichomes. In contrast, sheets of cells is formed from freeze-dried algae were not adhered in a linear pattern and no longer spherical.

2.4.2 Thermochemical Processing

This preparation includes both pyrolysis and hydrothermal liquefaction (HTL). Both procedures are damaging and ideal for generation of biofuels, yielding bio-oil, biochar and syngas. To avoid phase shift, hydrothermal liquefaction employs temperatures between 250 and 350 °C and moist algal biomass in pressured systems. The method yields both liquid and solid fractions. Biomass heated to 800 °C in pyrolysis heats anoxic air pressure, resulting in 3 stages in varying temperature. These fractions likely to be treated before being used as fuel. Yet, thermochemical alteration differs from standard lower-temperature bioprocesses, it is a very effective technique to extract the energy amount of biomass while maintaining a constant solid component. For example, during production of biodiesel from 13% extractable lipids containing microalgal biomass may necessitate drying, its extraction and desolventizing, yielding 130 kg of lipid content of dry microalgal biomass per tonne, HTL of wet microalgal biomass could yield a solvent fraction containing 400 kg of bio-oil, whereas direct, slower pyrolysis of the similar dry microalgal biomass could yield 270 kg of a crude oil content. From this, thermochemical processes convert not only lipid portion, but also proteins and carbohydrates, to bio-oil, resulting in a mixed composition of over 300 molecules. Very high temperatures pyrolysis at 800–1000 °C is a diversified gasification process. Both HTL and pyrolysis demand a lot of energy, yet the overall energy balance can be positive. Pyrolysis and HTL are projected to recover up to 78.7% and 89.9% of microalgae's energy amount, respectively.

Fractions of the procedure can be utilized to generate energy within for upcoming processing in a self-sufficient operation. High heating rates in Fast pyrolysis and short residence durations, whereas slow pyrolysis uses slower heating rates and longer residence times. HTL has residence periods similar to slow pyrolysis like one hour. HTL utilizes lower temperatures, yields higher liquid content conversions, and shows a better energy balance rather than pyrolysis. However, quick pyrolysis might also be appealing. The solid, gaseous and liquid fractions generated which are complicated and may necessitate upcoming processing such as neutralization, filtration and followed fractionation for specific applications [8].

2.4.3 Cell Disruption

This pretreatment stage includes numerous ways for permeating, disrupting, or dissolving cell membranes or walls. The numerous cell disruption strategies, which may be broadly classified as physical, chemical or mechanical. The cell membrane, lipid bilayer measuring 4 to 5 nm, serves as selective barrier that protecting internal components while also being easily thermally destroyed or dissolved. The algal cell wall is somewhat permeable to tiny molecules, missing in *D. salina*, but strong and thick in many microalgae, for example, 200 nm in *N. oleoabundans*. The mechanical disruption of the cell wall exposes organelles and internal macromolecular contents, and the hydrolysis is necessary to utilize cell wall components.

There is an essential contrast between physical approaches that are popular on a small scale and ways that are industrially scalable. Chemical processes are often scalable, although not all physical or mechanical procedures are. One of the criteria to evaluating process applicability is energy intensiveness, because the majority of energy employed in a mechanical disruption technique is converted into heat, limiting the application methods.

2.4.3.1 Sonication

Sonication is widely employed in laboratories for full biomass destruction, such as protein characterisation, nucleic acid extraction and lipid extraction. However, the procedure requires a lot of energy and may not be cost-effective. There are few scientific publications on large-scale sonication; most study reveals for its usage produced encouraging results on a small scale or used sonication in conjunction with extraction [8].

2.4.3.2 High Shear and Impaction

High-pressure homogenizers, microfluidization and impingement all operate on similar principles. The suspension is forced at higher speeds via a homogenizer valve, which experiences significant shear and decelerates. The system's strong velocity gradient causes cell deformation and disruption. These high velocities in tiny tubes demand high pressures, often 500–1500 bar.

2.4.3.3 Fluid Agitation

Fluid agitation is frequently insufficient to effectively destroy microalgae suspensions. Efficiency is boosted by using: (i) high velocity, (ii) abrasives or microbeads and (iii) precisely machined rotor–stator systems; in each condition, agitation creates significant shear and turbulence in the liquid biomass, and this process is energy demanding yet scalable. The precisely machined rotor–stator systems also makes dry grinding equipment, similar to jet mills [8].

2.4.3.4 Grinding of Powdered Biomass

Algal biomass that has been previously dried can be crushed in the same way as other biomasses or minerals are. The specific microalgal biomass, the use of grinding machines, and the desired ultimate size of the powder all influence the energy consumption and transfer to the product throughout this process. The energy input needs increasing with the contrary size of particle. Temperature increase is much crucial than in wet grinding; dry biomass rapidly warms to the denaturing temperatures except the operation is extended and lowered throughput. Freeze-thawing is a method that may be utilized for some microalgae, most notably *Arthrospira*, to produce phycocyanin and expose internal starch. In spite of the relatively large input of energy required for freezing, it is quite mild upon biomolecules [8].

2.4.3.5 Chemical Disruption Method

Alkali or Acid treatment, are often for small scale; nonetheless, an economic analysis is still required to assess the process. These processes overlay with targeted fractionation: full breakdown of algal cell walls with acid or alkali results in a carbohydrate-rich fraction, in addition to intracellular chemicals. Acid or basic hydrolysis is analogous to methods used for pretreatment of other algal biomasses, with some fractions, such as the cell membrane and hemicellulose, easily hydrolyzed and others, such as chitin or cellulose, resistant in milder temperatures. Membrane dissolving with solvents or detergents may be sufficient for cell penetration and upcoming processing.

2.4.3.6 Autolysis

Autolysis is a less-explored yet potentially effective method of cell destruction. It is dependent on species and can be induced by a variety of climatic stimuli, including anoxia when temperatures rise. Even though it is useful for a certain species, it is appealing because to the moderate circumstances and low-cost method, albeit slightly sluggish [8].

2.4.4 Targeted Fractionation or Solubilization

This pretreatment includes procedures that solubilize biomass components, comparable to cell disruption technique, but produce algal extracts that may then be fermented or purified. This phase may need a preparatory biomass disintegration process, and it is very varied due to the variety of molecular targets.

Targeted fractionation is the most important stage in producing high-value algal products such nutritious bio-oils and low-polarity components like xanthophylls, carotenoids, and vitamins. Another important feature of fractionation is transitioning from an ideal, targeted pretreatment to algal byproduct-generation stage. For example, rather than employing mechanical cell rupture to reveal cell composition, slower hydrolysis produces a fermentable content. The fractionation stage is slower, even though it has the potential to increase the overall profitability of the biorefinery process. The removal of lipids can be favorable for protein-rich fractions. In *Arthrospira* sp., the lipid fractions and cell membrane dissolve easily, resulting in a desirable combination of components with low polarity. The following disintegration of proteins in appropriate buffers yields another percentage of discarded algal biomass that may be processed into biodigested, bio-oils, or fed.

References

1. Baicha Z, Salar-García MJ, Ortiz-Martínez VM,Hernández-Fernández FJ, de los Ríos AP, Labjar N, Lotfi E, Elmahi M (2016) A critical review on microalgae as an alternative source for

bioenergy production: A promising low cost substrate for microbial fuel cells. Fuel Processing Technology, 154:104. https://doi.org/10.1016/j.fuproc.2016.08.017.

2. Barros A, Pereira H, Campos J, Marques A, Varela J, Silva J (2019) Heterotrophy as a tool to overcome the long and costly autotrophic scale-up process for large scale production of microalgae. Scientific Reports, 9(1):13935.
3. Barros AI, Gonçalves AL, Simões M, Pires JC (2015) Harvesting techniques applied to microalgae: a review. Renewable and sustainable energy reviews, 41, 1489–1500.
4. Borowitzka MA (1999) Commercial production of microalgae: ponds, tanks, tubes and fermenters. Journal of Biotechnology, 70(1–3):313–321.
5. Castillo T, Ramos D, García-Beltrán T, Brito-Bazan M, Galindo E (2021) Mixotrophic cultivation of microalgae: an alternative to produce high-value metabolites. Biochemical Engineering Journal, 176:108183. https://doi.org/10.1016/j.bej.2021.108183.
6. Chamkalani A (2018) Dynamic thermodynamic flux balance analysis and life cycle analysis of microbial biofuels (Doctoral dissertation, Memorial University of Newfoundland).
7. Chisti Y (2007) Biodiesel from microalgae. Biotechnology Advances, 25(3), 294-306.
8. de Carvalho JC, Magalhães Jr AI, de Melo Pereira GV, Medeiros AB, Sydney EB, Rodrigues C, Aulestia DT, de Souza Vandenberghe LP, Soccol VT, Soccol CR (2020) Microalgal biomass pretreatment for integrated processing into biofuels, food, and feed. Bioresource Technology, 300:122719. https://doi.org/10.1016/j.biortech.2019.122719.
9. Deepa P, Sowndhararajan K, Kim S (2023) A review of the harvesting techniques of microalgae. Water, 15(17):3074. https://doi.org/10.3390/w15173074
10. Dos Santos LB, Bassin ID, Cammarota MC (2023) A comparative analysis of algal biomass densification methods aimed at the biofuel production. World Journal of Advanced Engineering Technology and Sciences, 08(01):132–156 https://doi.org/10.30574/wjaets.2023.8.1.0022.
11. Dutta S, Kataki S, Banerjee I, Pohrmen CB, Jaiswal KK, Jaiswal AK (2025) Microalgal biorefineries in sustainable biofuel production and other high-value products. New Biotechnology. https://doi.org/10.1016/j.nbt.2025.02.007.
12. Figueira Garcia L, Gojkovic Z, Venuleo M, Guidi F, Portillo E (2024). The Use of Chemical Flocculants and Chitosan as a Pre-Concentration Step in the Harvesting Process of Three Native Microalgae Species from the Canary Islands Cultivated Outdoors at the Pilot Scale. Microorganisms, 12(12):2583. https://doi.org/10.3390/microorganisms12122583.
13. Gupta SK, Bux F (2019) Application of microalgae in wastewater. Springer Nature Switzerland AG, Domestic and industrial wastewater treatment. https://doi.org/10.1007/978-3-030-13913-1.
14. Jacob-Lopes E, Queiroz MI, Zepka LQ (2020) Pigments from Microalgae Handbook; Springer International Publishing: Cham, Switzerland.
15. Johnson TJ (2016) Enhancing the industrial potential of filamentous cyanobacteria. South Dakota State University.
16. Krujatz F, Illing R, Krautwer T, Liao J, Helbig K, Goy K, Opitz J, Cuniberti G, Bley T, Weber J (2015). Light-field-characterization in a continuous hydrogen-producing photobioreactor by optical simulation and computational fluid dynamics. Biotechnology and Bioengineering, 112(12):2439–2449. https://doi.org/10.1002/bit.25667.
17. Mata TM, Martins AA, Caetano NS (2010) Microalgae for biodiesel production and other applications: a review. Renewable and Sustainable Energy Reviews, 14(1):217–232.
18. Parray JA, Singh N, Haghi AK (2024) Microalgae as a Sustainable Source of Green Energy Generation and Bioeconomy. Springer Nature. https://doi.org/10.1007/978-3-031-77703-5.
19. Reddy SD, Deepika N, Dropathi MR, Vishwanutha S, Daaman JD, Reddy CN, Yadavalli R (2023) Factors affecting the microalgal biomass productivity in photobioreactors. In Current Developments in Biotechnology and Bioengineering (pp. 59–88). Elsevier. https://doi.org/10.1016/B978-0-323-99911-3.00003-8.

20. Richmond A. (2004) Handbook of microalgal culture: biotechnology and applied phycology. Blackwell Science.
21. Sarangi PK (Ed.) (2023) Biorefinery Production of Fuels and Platform Chemicals. John Wiley & Sons. https://doi.org/10.1002/9781119724872.
22. Sharma S, Sahu S, Arya SK. (2024) Pretreatment and Fractionation of Algae Biomass for Value-Added Extraction. In Value Added Products From Bioalgae Based Biorefineries: Opportunities and Challenges (pp. 95–115). Singapore: Springer Nature Singapore. https://doi.org/10.1007/978-981-97-1662-3_5.
23. Udayan A, Sirohi R, Sreekumar N, Sang BI, Sim SJ (2022) Mass cultivation and harvesting of microalgal biomass: Current trends and future perspectives. Bioresource technology, 344:126406. https://doi.org/10.1016/j.biortech.2021.126406.
24. Ugwu CU, Aoyagi H, Uchiyama H (2008) Photobioreactors for mass cultivation of algae. Bioresource Technology, 99(10):4021–4028. (actabiologicaturcica.com)
25. Upadhyay M, Hansa A, Devi A, Lajayer BA, Price GW, Sharma R (2024) Status and future scope of algal biomass-based remediation for various environmental contaminants. Journal of Water Process Engineering, 65:105809. https://doi.org/10.1016/j.jwpe.2024.105809.
26. Wang JH, Zhuang LL, Xu XQ, Deantes-Espinosa VM, Wang XX, Hu HY (2018) Microalgal attachment and attached systems for biomass production and wastewater treatment. Renewable and Sustainable Energy Reviews, 92:331–342. https://doi.org/10.1016/j.rser.2018.04.081.
27. Wang M, Ye X, Bi H, Shen Z (2024) Microalgae biofuels: illuminating the path to a sustainable future amidst challenges and opportunities. Biotechnology for Biofuels and Bioproducts, 17(1):10. https://doi.org/10.1186/s13068-024-02461-0.
28. Wo WM, Hamza AZ (2024) Potential of marine microalgae for the production of food and feed industry. Marine biomass: Biorefinery, bioproducts and environmental bioremediation, 249–278. https://doi.org/10.1515/9783111353951-011.

Biomass Conversion Processes 3

Abstract

This chapter examines different biomass conversion processes utilised in the production of bioenergy from microalgae. The text addresses thermochemical conversion, biochemical conversion, hydrothermal liquefaction (HTL), and biogas production via anaerobic digestion. Each process is analysed comprehensively, emphasising its benefits, constraints, and possible applications. This chapter offers an overview of conversion technologies, facilitating an understanding of the complexities and opportunities associated with microalgal biomass conversion. This chapter examines recent advancements and innovations in bioenergy production processes, aiming to enhance the sustainability and efficiency of microalgal biomass utilisation.

3.1 Introduction

The process of converting biomass into useful chemicals is rapidly evolving. Biomass resources must be converted into valuable products, such as fuels and chemicals, for use in both domestic and commercial applications. Biomass conversion techniques may employ biochemical, physiochemical, or thermochemical reactions (Fig. 3.1).

3.2 Thermochemical Processes

Garba [22] and Matsumura [45] indicate that biomass can be converted into biochar, bio-oil, natural gas, heat, and both thermal and electrical energy via various thermochemical processes. In the thermochemical conversion process, parts of organic biomass are heated up and broken down into chemicals that can be used. Perera et al. [52] say

H. N. Akolkar et al., *Bioenergy from Microalgal Biomass*, Synthesis Lectures on Renewable Energy Technologies, https://doi.org/10.1007/978-3-032-03434-2_3

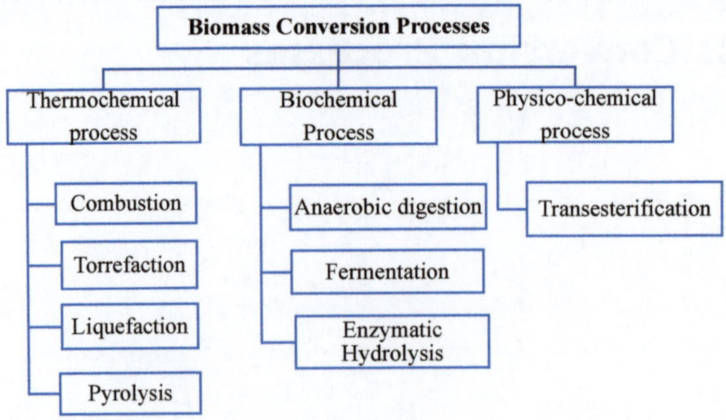

Fig. 3.1 Classification of biomass conversion processes

that this method can be used instead of controlling the temperature or oxidation of biomass to make biofuel. Garba [22] and Matsumura [45] identify several factors that affect the selection of conversion technology, including biomass accessibility, biomass type and quantity, cost-effectiveness of the product options, manufacturing process, and environmental considerations.

The biomass feedstock needs to be low in moisture and solid. Similarly, using inappropriate technology can result in decreased productivity [27, 80]. Pyrolysis, torrefaction, gasification, hydrothermal liquefaction (HTL), and combustion are frequently included in these thermochemical processes [28] Fig. 3.2.

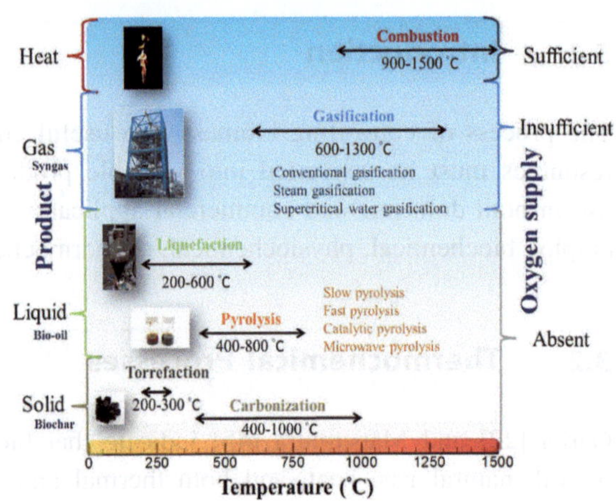

Fig. 3.2 A schematic diagram of thermochemical conversion processes of biomass. *Source* Chen et al. [12]

3.2.1 Pyrolysis

Pyrolysis is a process that converts biomass into energy by heating it at high temperatures and specific pressures in an environment with minimal to no oxygen, rather than burning it. Pyrolysis is a flexible and effective technique that makes it possible to produce power, heat, and chemicals efficiently by converting solid biomass into a liquid that is convenient to transport and store. Shredded beef and sludge represent high-moisture waste biomass requiring drying prior to pyrolysis [75].

For the pyrolysis of wood, this complicated method uses both sequential and simultaneous steps. When biomass is heated in an inert atmosphere, the components begin to thermally decompose between 350 to 550 °C, which speeds up to 700 to 800 °C when air is absent. Pyrolysis is the process by which organic matter (biomass) is decomposed into gases and condensable vapours (tars, oils, solid charcoal, etc.), the byproducts of which are smaller molecules containing carbon, hydrogen, and oxygen.

Depending on the pyrolysis process settings, each of these components decomposes to varying degrees and at varying rates [21, 32].

Among the subcategories of pyrolysis are flash pyrolysis, catalytic hydropyrolysis, quick pyrolysis, slow pyrolysis, fast pyrolysis, and catalytic pyrolysis [8]. When heat is provided from an external energy source—often the incomplete combustion of the biomass feedstock or the combustion of the generated gases—slow pyrolysis typically takes place at atmospheric pressure. In these conditions, biochar production are frequently low [38]. Furthermore, fast pyrolysis is the quickest way to turn biomass into gases, liquid fuel, and bio-oil. This technique usually requires an inert atmosphere and a mild temperature. The efficiency of the process and the calibre of the end product are impacted by additional operating variables, though, including temperature, catalysts, pressure, additives, and reactor type [30]. Lastly, low-quality biomass energy may be converted into high-grade biomass energy using a cutting-edge technique known as flash pyrolysis [79].

Biomass flash pyrolysis frequently generates a bio-oil yield of 70–85 weight percent. The oxygenated compounds that constitute it are highly variable in composition, contingent upon the biomass type and operating conditions [2]. This process is characterised by a brief residence duration (often less than 0.5 s) and rapid heating rates (103–104 °C/s), as per Kan et al. [34].

Since they are superior to the conventional techniques of assisting biomass pyrolysis with heat or electricity, two cutting-edge pyrolysis technologies—microwave aided pyrolysis and solar-assisted pyrolysis—have attracted a lot of attention recently [14, 62]. According to Chintala et al. [13], one technique used a solar concentrator to gather solar energy, which was then used to power biomass pyrolysis, while the other employed agitating molecules in an electromagnetic field to transform microwave energy into heat.

3.2.2 Combustion

Combustion is a series of reactions that release heat, or exothermic energy. Energy is released when chemical bonds break, which is a common function of these bonds. The majority of companies and power plants use this energy to create steam, which is then used by turbines to generate electricity and heat. When discussing alternative fuels like biomass, the term "combustion" refers to the process of burning organic materials. According to Abuelnuor et al. [1], wood is the fuel of choice for biomass combustion. From a few kilowatts (kW) to over a hundred gigawatts (GW), biomass combustion systems cover a wide size spectrum. It is possible to generate heat from biomass in an extremely cost-effective and efficient manner. Steam plants' particular cost and efficiency are most noticeable in applications at huge sizes [50]. Macroscopic and microscopic models are the usual categories for biomass combustion. In particular, the macroscopic characteristics of biomass, including particle size, heating value, moisture content, density, and ash fusion temperature, are given for macroscopic study. Among the many characteristics that can be studied by microscopic examination are thermal and chemical kinetics as well as mineral data [16]. Fuel, energy, and environmental factors may be considered to interact during the three fundamental combustion operations. The combustion of biomass fuel in the boiler results in the release of vapours that are combustible and can be ignited by flames. When more air is introduced, the residual substance, which is still carbon char, will eventually catch fire. Combustion heat can be utilised in subsequent conversion processes to produce electrical energy; however, this is conditional on a number of other factors [69].

3.2.3 Torrefaction

Torrefaction is a thermal process that turns biomass into a charred product that can be utilised as organic manure or as fuel. Torrefied substances are often called biocoal when utilised as fuel. Conversely, the term "biochar" is predominantly used when the torrefied product is utilised as a soil amendment [5]. Torrefaction is a thermal pre-treatment method that contributes to the reduction of biomass heterogeneity in terms of its chemical and physical properties. It reduces the moisture content, increases its energy content, and lessens biological degradation [74]. The typical temperature range for the torrefaction process is between 225 and 300 °C. During torrefaction, wood's lignin and hemicelluloses are partially broken down. The pelletisation of the product can result in a density that exceeds that of conventional pellets, while torrefaction enhances its calorific value [10]. For a biomass torrefaction plant to be economically sustainable, the energy released in the volatiles must be used.

In order to handle significant levels of moisture and help produce the energy required for the drying process, additional backup fuel is needed. The degree of torrefaction

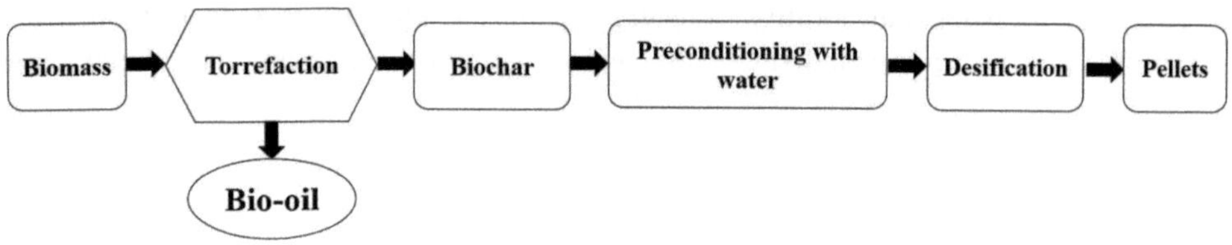

Fig. 3.3 Torrefaction process

depends on how long the biomass is exposed to it and the temperature at which it happens [7]. Biochar's poor mechanical strength can be strengthened via densification, potentially halting a large mass loss during handling and transit. According to recent research, biochar's poor mechanical qualities, which result from the material's densification, limit its use in a number of potential applications [63] (Fig. 3.3).

3.2.4 Liquefaction

A potential thermochemical conversion method called biomass liquefaction turns biomass into biochar (10–12%), gases (35–45%), and bio-oil (target product, 40–60%). It is frequently carried out at a high pressure (5–25 MPa) and relatively low temperature (200–600 °C) in a solvent, such as ethanol, glycerol, acetone, water, phenols, and tetralin [9]. Process variables, including biomass type, reaction pressure, temperature, residence time, catalyst, solvent, etc., are the main determinants of the product compositions and ensuing characteristics [77]. Process parameter optimisation is required to produce bio-oil in an efficient and high-quality manner. Biomass with higher levels of cellulose and hemicellulose may produce more bio-oil [78]. It was easier to liquefy cellulose and hemicellulose due to their low breakdown temperatures and simple structures. One important factor in biomass liquefaction is the reaction temperature, which has a big impact on the calorific values and compositions of bio-oils.

To boost conversion rates, biomass liquefaction uses high pressure to allow solvents to penetrate the biomass, which aids in the destruction of the macromolecules [59]. The impact of pressure (200–400 bar) on the liquefaction of biomass was investigated by Basar et al. [6]. At pressures greater than 221 bar, they discovered that the bio-oil output was unaffected by the pressure. Moreover, high temperatures often accompany high pressures, which typically leads to negative effects beyond critical values. Decomposition of biomass and synthesis of bio-oil are both aided by an appropriately long retention period. The condensation processes and bio-oil cracking are both accelerated by an excessively long retention period, but the bio-oil yield is decreased since the intermediate products re-polymerize to coke or petrol.

The high pressure and additive consumption during biomass liquefaction limits its industrial scale utilisation and raises the reactor system's complexity, challenges, and costs [20].

3.2.5 Gasification

According to Ren et al. [61], gasification is an endothermic process that involves heating biomass to higher temperatures (600–1300 °C) with gasification agents such steam, carbon dioxide, and air/O_2. Hydrocarbons, CO, CO_2, H_2, and CH_4 are the gaseous products. Due to its ability to mitigate the negative effects of biomass's high O levels, steam is the ideal gasification agent for increasing gasification products' calorific values and improving their compositions [76] (Fig. 3.4).

Typically, in the biomass gasification process, interactions between gasifying agents and char serve as the controlling phase owing to their comparatively sluggish reaction speeds. The char's reactivity is significantly influenced by volatile components, including hydrocarbons, tarry substances, steam, and other light gaseous species [82]. The fundamental framework of the combustion process is defined below. Sikarwar et al. [67] identified three main gasifier types used in biomass gasification: fluidized-bed, entrained-flow, and fixed-bed gasifiers. There are two main kinds of fixed bed gasifiers: updraft and downdraft (Fig. 3.5).

Biomass is injected from the top and descends in an updraft gasifier, while gasifying ingredients are supplied from the bottom, resulting in the upward movement of the generated gas. In this scenario, combustion transpires at the gasifier's most elevated temperature zone, the base of the bed, while the resulting gas is expelled from the top at a reduced temperature. Because of the lower exit temperature, the resulting gas contains a lot of tar. The downdraft gasifier takes in both the feed gas and the product gas, and the latter leaves

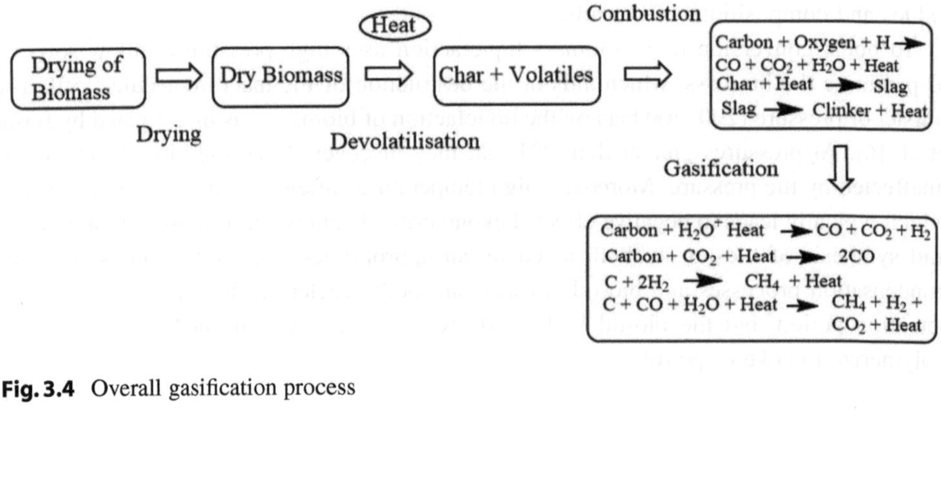

Fig. 3.4 Overall gasification process

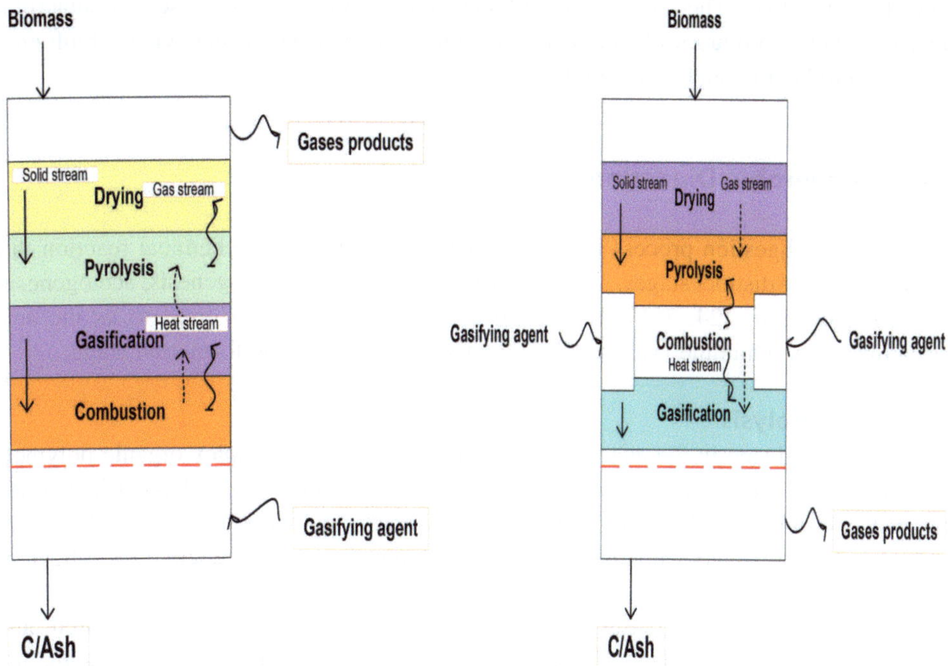

Fig. 3.5 Fixed bed gasifiers: updraft and downdraft gasifiers. *Source* Joshi et al. [33]

the former at a higher temperature. Gas flowing from a hot area incinerates most of the tar in this situation [36]. Suspended biomass particles are exposed to an oxygen-enriched environment in fluidized-bed gasifiers. The resultant bed exhibits fluid-like behaviour. In order to combine biomass with coal that has already been gasified, some gasifiers use a process called back-mixing. While the oxidant and steam are delivered from the bottom at a sufficient velocity, biomass particles are introduced from the reactor side. According to Ram and Mondal [56], these gasifiers usually operate at high temperatures in order to breakdown the byproducts that accumulate at the periphery and to achieve a high carbon conversion rate. Cross-draft gasification is a straightforward form of gasification that employs a reactor similar to an updraft gasifier, where the fuel is introduced from the top and experiences a slow thermochemical reaction as it falls within the reactor [64].

3.3 Biochemical Processes

Biochemical conversion relies on the utilization of microbes or enzymes to transform biomass into biofuel [29]. The biochemical conversion procedure may be appropriate for wet algal biomass and is typically appropriate for organic wastes with a high moisture content (>50%) [51]. The most common biochemical conversion processes are anaerobic digestion, which produces alcoholic fermentation, biomethane, which yields biological hydrogen production, and bioethanol.

3.3.1 Anaerobic Digestion (AD)

The anaerobic digestion process is a comprehensively understood chemical reaction that transpires in four distinct stages. The processes are hydrolysis, acidogenesis, acetogenesis, and methanogenesis [24, 35]. The second and third stages are referred to as the acid formation stage, while the last stage is designated as the methane formation stage.

3.3.1.1 Hydrolysis

This is the initial step in the anaerobic digestion process. The complex organic polymers are transformed into simple, soluble molecules at this point. In this phase, proteins are changed into amino acids, lipids (fats) into fatty acids, and carbohydrates (polysaccharides) into simple sugars (monosaccharides). Different types of facultative or obligatory fermentative bacteria use extracellular enzymes to carry out the hydrolysis stage. Cellulases, amylases, and pectinases hydrolysed polysaccharides like cellulose, starch, and pectin to monosaccharides, while lipases change lipids into long-chain fatty acids and proteases change proteins into amino acids [11].

$$(C_6H_{10}O_5)_n + nH_2O = n(C_6H_{12}O_6)$$

3.3.1.2 Acidogenesis

The second stage involves the conversion of the soluble compounds generated by hydrolysis into hydrogen, carbon dioxide, ethanol, volatile fatty acids (C1–C5), and certain organic nitrogen and sulfur compounds. Acetic acid (CH_3COOH), butyric acid ($CH_3CH_2CH_2COOH$), propionic acid (CH_3CH_2COOH), and valeric acid ($CH_3CH_2CH_2CH_2COOH$) are the acids generated at this stage. The other products are sent to the third stage for additional acetogen breakdown, whereas the acetic acid produced in this stage is taken straight to the last stage.

$$C_6H_{12}O_6 + H_2O \rightarrow 2CH_3COOH + 4H_2 + CO_2$$

$$C_6H_{12}O_6 + 2H_2 \rightarrow 2CH_3CH_2COOH + 2H_2O$$

$$C_6H_{12}O_6 \rightarrow CH_3CH_2CH_2COOH + 2H_2 + 2CO_2$$

$$C_6H_{12}O_6 \rightarrow 2CH_3CH_2OH + 2CO_2$$

$$C_6H_{12}O_6 \rightarrow 2CH_3CHOHCOOH$$

3.3.1.3 Acetogenesis

Acetogenesis is the third step in the anaerobic digestion process. During this phase, acetogens help transform volatile fatty acids containing more than two carbon atoms (from the acidogenesis step) into acetic acids, carbon dioxide, and hydrogen [35].

$$CH_3CH_2OH + H_2O \rightarrow 2CH_3COOH + 2H_2$$

$$2CH_3CH_2OH + 2CO_2 \rightarrow CH_4 + 2CH_3COOH$$

$$CH_3CH_2COOH + 2H_2O \rightarrow CH_3COOH + 3H_2 + CO_2$$

$$CH_3CH_2CH_2COOH + 2H_2O \rightarrow 2CH_3COOH + 2H_2$$

$$CH_3CHOHCOOH + H_2O \rightarrow CH_3COOH + CO_2 + 2H_2$$

3.3.1.4 Methanogenesis

In this final phase, acetic acid, hydrogen, and a little amount of carbon dioxide are consumed by methanogenic bacteria, or methogens, to generate methane. According to Nayono [49], the process of acetic acid decarboxylation yields around 66% methane, while the reduction of carbon dioxide accounts for the remaining 34%.

$$CH_3COOH \rightarrow CH_4 + CO_2$$

$$CO_2 + 4H_2 \rightarrow CH_4 + 2H_2O$$

3.3.1.5 Fermentation

Fermentation facilitates the transformation of biomass into valuable bio-products (such as biogas, biohydrogen, and bioethanol) by the action of microorganisms.

There are two types of fermentation: photo fermentation and dark fermentation. According to Sung et al. [72], the photofermentation (PF) process involves the use of light energy by photosynthetic bacteria, namely purple non-sulfur (PNS) bacteria, to transform organic compounds such as butyrate, lactate, and acetate into carbon dioxide and biohydrogen. The type of substrate utilized in the PF process can affect the ratio of carbon dioxide to biohydrogen produced [81]. Nitrogen-limited circumstances are necessary for the evolution of biohydrogen because PNS bacteria use the nitrogenase enzyme to produce and fix nitrogen at the same time. In order to produce electrons and adenosine triphosphate (ATP), which in turn propel nitrogenase to evolve biohydrogen and fix molecular nitrogen into ammonium ions, organic acids must first be broken down using light energy [3]. The purple non-sulfur bacterial species Rhodobacter sphaeroides, Rhodopseudomonas palustris, Rhodospirillum rubrum, Rhodobacter capsulats, and Rubrivivax gelatinosus are frequently employed for the synthesis of biohydrogen. Microalgae biomass must first undergo hydrolysis since the PNS bacteria participating in the PF process can only eat tiny organic compounds, typically organic acids. Thus, dark fermentation probably comes before the PF process [18, 53].

The steps of hydrolysis, acidogenesis, and acetogenesis that organic substrates go through in the dark fermentation (DF) process are comparable to AD process. But during DF, the methanogenic step is inhibited. During the acidogenesis and acetogenesis phases of the AD process, biohydrogen is usually generated as an intermediate metabolite. Methanotrophs that use hydrogen then convert it to biomethane. To stop the fermentation process at the acetogenesis-acidogenesis stage, more work is done during DF than in AD to inactivate methanotrophs from the original inoculum [25]. Like AD, DF's main problem is the microalgal biomass's poor biodegradability because of its resistance to cell walls [48]. The proper pretreatment method must be used before fermentation in order to break down the cell wall and convert the complex carbohydrates into simpler monosaccharides. When PF is introduced, the low biohydrogen generation efficiency of DF can be significantly raised. According to published studies [18, 53], DF not only produces biohydrogen but also hydrogenic effluent that contains a variety of organic acids that PF can use to make more biohydrogen. The need to construct specialised anaerobic photobioreactors with sizable surface surfaces exposed to sunlight, however, impedes the combined DF–PF process and raises the system's complexity and expense [47].

3.4 Physiochemical Processes

3.4.1 Transesterification

When a catalyst is present, monoalcohols interact with microalgae lipids (triglycerides/free fatty acids) in the transesterification process, which yields biodiesel [60]. In order to move the equilibrium to the right and enable higher response rates, these equilibrium mechanisms require a high alcohol/bio-oil ratio [68]. Depending on whether an acid, base, or enzyme catalyst is used, the transesterification reaction can be divided into three groups. Each of the three categories of transesterification processes can be further classified into two broad categories: homogeneous transesterification and heterogeneous transesterification. Homogeneous transesterification reactions are favoured because of the catalyst's inexpensive cost and the rapid reaction rate. Conversely, in heterogeneous transesterification reactions, the separation and regeneration of the catalyst is far more facile, thereby enabling its application on a greater scale [40]. The enzymatic transesterification reaction is a prominent study area because to advantages including facile product separation, absence of saponification, and operation at significantly lower temperatures; however, commercialization requires adjustment of the operational parameters [15].

3.4.1.1 Base-Catalyzed Transesterification

Using basic homogeneous or heterogeneous catalysts to transesterify lipids is the most popular way to produce biodiesel [73]. Comparing these reactions to those that are

Table 3.1 Production of biodiesel through basic homogeneous transesterification

Basic catalysts	Feed	Operating parameters	Mono-alcohol	FAC [%]	References
KOH	Rapeseed	65 °C, 2 h	Methanol	91	Rashid and Anwar [57]
	Frying oil	65 °C, 2 h		92	Encinar et al. [19]
NaOH	Sunflower	60 °C, 2 h		90	Rashid et al. [58]
	Frying oil	65 °C, 1.5 h		92	Encinar et al. [19]
	Mahua (Madhuca indica)	60 °C, 2 h		92	Puhan et al. [55]
RbOH	Sunflower oil	50 °C, 2 h		90	Arzamendi et al. [4]
LiOH	Sunflower oil	50 °C, 2 h		90	Arzamendi et al. [4]

FAC: Fatty Acid Conversion

catalysed by enzymes and acids, they are the fastest [15]. The heterogeneous basic transesterification process is environmentally friendly, characterised by low separation costs and straightforward catalyst generation, facilitating scalability. The homogeneous transesterification reaction uses both potassium and sodium hydroxides. The commercial synthesis of biodiesel makes substantial use of these two bases [37, 65]. Table 3.1 presents a summary of the latest research conducted on the use of homogeneous base catalysts in the production of biodiesel, specifically fatty acid methyl esters (FAME). For example, 85% of the FAME production can be achieved in 0.5 h by converting waste oil into biodiesel over homogeneous sodium hydroxide and methyl alcohol [39].

A summary of recent research on transesterification processes using heterogeneous base catalysts may be found in Table 3.2. Heterogeneous base catalyst-induced transesterification reactions are significantly more important in industry than homogeneous catalyst-facilitated reactions for a number of reasons, including the fact that the solid catalyst can be economically separated from the liquid effluent without emulsion formation or corrosion occurring and the fact that there is no need to worry about environmental contaminants [41, 70].

3.4.1.2 Acid-Catalyzed Transesterification

Either base- or acid-catalyzed transesterification processes use the same reversible stages in their reaction pathways. In most cases, an acidic transesterification reaction can achieve

Table 3.2 Production of biodiesel through basic heterogeneous transesterification

Basic catalysts	Feed	Operating parameters	Mono-alcohol	FAC [%]	References
CaO	Sunflower oil	252 °C, 6 min	Methanol	99	Demirbas [17]
KF/CaO	*Chlorella vulgaris*	65 °C, 45 min		93.07	Ma et al. [43]
K_3PO_4	Cooking oil	50 °C, 90 min		92	Madhuvilakku and Piraman [44]
TiO_2-ZnO	Palm oil	65 °C, 5 h		92.2	Madhuvilakku and Piraman [44]
KF/Eu_2O	Rapeseed	65 °C, 1 h		92	Sun et al. [71]

a FAME yield of 95–99%; however, a somewhat larger alcohol-to-oil ratio (usually 9:1–166:1) is required to tip the balance in favour of the product side [46]. The substantial quantity of mono-alcohol complicates the separation process, whereas a reduced concentration of alcohol in the feed necessitates an increased reaction temperature to achieve optimal product yield. A higher molar ratio of alcohol to oil is typically employed, as utilising the lowest feasible reaction temperature for lipid transesterification minimises selectivity towards undesired products and reduces material corrosion [15].

Unlike basic transesterification processes, acid-catalyzed reactions simultaneously esterify and transesterify, therefore there is no possibility of soap production despite the feed oil's increased acidity. Table 3.3 summarises some noteworthy results from the homogeneous acid-catalyzed transesterification of microbial oil using sulphuric acid (H_2SO_4) and hydrochloric acid (HCl), the two most widely utilised catalysts.

3.4.1.3 Enzyme-Catalyzed Transesterification

One promising alternative to traditional methods of basic and acidic transesterification is the use of lipases to convert microbial oil into biodiesel. This process is both cost-effective and eco-friendly. The advantageous motivations for enzyme-assisted transesterification reactions include the production of high-purity biodiesel, absence of saponification, reduced operational costs, and straightforward separation and recycling processes. Consequently, enzyme-catalyzed transesterification for biodiesel generation represents the optimal sustainable alternative to petro-diesel [54]. In typical practice, lipases are employed in both free-form and immobilised forms for the purpose of homogenous transesterification.

The three different methods for transforming microbial oil into biodiesel—base-, acid-, and enzyme-catalyzed transesterification—are thoroughly compared in Table 3.4.

Table 3.3 Production of biodiesel through acid catalysed transesterification

Acid catalysts	Feed	Operating parameters	Mono-alcohol	FAC [%]	References
Homogeneous					
H_2SO_4 (0.5 wt %)	Soybean	100 °C, 12 h	Methanol	98	Goff et al. [26]
H_2SO_4 (15 wt %)	*Jatropha curcas L.* seeds	60 °C, 24 h	Methanol/ n-hexane	99.8	Shuit et al. [66]
H_2SO_4 (0.5 M)	*Chlorella pyrenoidosa*	110 °C, 2 h	Methanol/ n-hexane	95	Li et al. [42]
Heterogeneous					
ZS/Si	Cooking oil	200 °C, 5 h	Methanol	98	Jacobson et al. [31]
S-ZrO_2	Soybean	120 °C, 1 h	Methanol	98.6	Garcia et al. [23]

Table 3.4 Comparison of transesterification processes [15]

Process characteristics	Acid-catalysed transesterification	Base-catalysed transesterification	Enzyme-catalysed transesterification
FAME yield	High yield (>94%)	High yield (>94%)	High yield (>90%)
Alcohol/oil	Excess	Moderate	Stoichiometric ratio
Temperature	Elevated (60–120 °C)	Moderate (50–70 °C)	Low (30–60 °C)
Catalyst	Inexpensive	Inexpensive	Expensive
Reaction time	Long (~4 h)	Short (~12 h)	Long (~28 h)
Product recovery	Easy	Difficult	Easy
Saponification	Transformation of FFA without soap	Soap formation	Transformation of FFA without soap

References

1. Abuelnuor AAA, Wahid MA, Hosseini SE, Saat A, Saqr KM, Sait HH, Osman M (2014) Characteristics of biomass in flameless combustion: a review. Renew Sustain Energy Rev 33: 363–370. https://doi.org/10.1016/j.rser.2014.01.079.
2. Amutio M, Lopez G, Alvarez J, Moreira R, Duarte G, Nunes J, Bilbao J (2013) Flash pyrolysis of forestry residues from the Portuguese Central Inland Region within the framework of the BioREFINA-Ter project. Bioresour Technol 129:512–518. https://doi.org/10.1016/j.biortech.2012.11.114.

3. Arimbrathodi SP, Javed MA, Hamouda MA, Aly Hassan A, Ahmed ME (2023) BioH_2 production using microalgae: Highlights on recent advancements from a bibliometric analysis. Water 2023, 15:185. https://doi.org/10.3390/w15010185.
4. Arzamendi G, Arguiñarena E, Campo I, Zabala S, Gandía LM (2008) Alkaline and alkaline-earth metals compounds as catalysts for the methanolysis of sunflower oil. Catal Today 133:305. https://doi.org/10.1016/j.cattod.2007.11.029.
5. Barskov S, Zappi M, Buchireddy P, Dufreche S, Guillory J, Gang D, Sharp R (2019) Torrefaction of biomass: a review of production methods for biocoal from cultured and waste lignocellulosic feedstocks. Renew. Energy 142:624–642. https://doi.org/10.1016/j.renene.2019.04.068.
6. Basar IA, Liu H, Carrere H, Trablyb E, Eskicioglu C (2021) A review on key design and operational parameters to optimize and develop hydrothermal liquefaction of biomass for biorefinery applications. Green Chem 23:1404–46. https://doi.org/10.1039/D0GC04092D
7. Basu P (2018) Biomass gasification, Pyrolysis and torrefaction: Practical Design and Theory. Academic Press.
8. Bhatnagar P, Gururani P, Bisht B, Kumar V (2021) Algal Biochar: an advance and sustainable method for wastewater treatment. Octa J Biosci 9 (2).
9. Bodur FG, Güngören-Madenoğlu T, Özdemir G, Ballice L, Kabay N (2023) Biochemical and biohythane production from anaerobically digested water plant by hydrothermal liquefaction/gasification. Int J Hydrogen Energy https://doi.org/10.1016/j.ijhydene.2023.03.455.
10. Caillat S, Vakkilainen E (2013) Large-scale biomass combustion plants: an overview. Biomass Combus. Sci. Technol. Eng. 189–224. https://doi.org/10.1533/9780857097439.3.189.
11. Chanakya HN and Malayil S (2012) Anaerobic digestion for bioenergy from agro- residues and other solid wastes-an overview of science, technology and sustainability, J Indian Inst Sci., 92:111–143.
12. Chen WH, Lin BJ, Lin YY, Chu YS, Ubando AT, Show PL, et al. (2021) Progress in biomass torrefaction: principles, applications and challenges. Prog Energy Combust Sci 82:100887. https://doi.org/10.1016/j.pecs.2020.100887.
13. Chintala V, Kumar S, Pandey JK, Sharma AK, Kumar S (2017) Solar thermal pyrolysis of non-edible seeds to biofuels and their feasibility assessment. Energy Convers Manag 153:482–92. https://doi.org/10.1016/j.enconman.2017.10.029.
14. da Silveira Rossi RA, Barbosa JM, de Souza Barrozo MA, Martins Vieira LG (2021) Solar assisted catalytic thermochemical processes: pyrolysis and hydropyrolysis of Chlamydomonas reinhardtii microalgae. Renew Energy 170:669–82. https://doi.org/10.1016/j.renene.2021.02.034.
15. del Pilar Rodriguez M, Brzezinski R, Faucheux N, Heitz M, (2016) Enzymatic transesterification of lipids from microalgae into biodiesel: a review. AIMS Energy 2016, 4 (6):817. https://doi.org/10.3934/energy.2016.6.817.
16. Demirbas A (2004) Combustion characteristics of different biomass fuels. Prog Energy Combust Sci 30(2):219–230. https://doi.org/10.1016/j.pecs.2003.10.004.
17. Demirbas A (2007) Biodiesel from Sunflower Oil in Supercritical Methanol with Calcium Oxide. Energy Conversion and Management, 48:937–941. https://doi.org/10.1016/j.enconman.2006.08.004.
18. Ding L, Cheng J, Lu H, Yue L, Zhou J, Cen K (2017) Three-stage gaseous biofuel production combining dark hydrogen, photo hydrogen, and methane fermentation using wet Arthrospira platensis cultivated under high CO_2 and sodium stress. Energy Convers Manag 148: 394–404. https://doi.org/10.1016/j.enconman.2017.05.079.

19. Encinar JM, Gonzâlez JF, Rodríguez-Reinares A (2005) Biodiesel from Used Frying Oil. Variables Affecting the Yields and Characteristics of the Biodiesel. Ind Eng Chem Res 44(15):5491. https://doi.org/10.1021/ie040214f.
20. Fan YJ, Hornung U, Dahmen N (2022) Hydrothermal liquefaction of sewage sludge for biofuel application: a review on fundamentals, current challenges and strategies. Biomass Bioenergy 165:106570. https://doi.org/10.1016/j.biombioe.2022.106570.
21. Fisher T, Hajaligol M, Waymack B, Kellogg D (2002) Pyrolysis behavior and kinetics of biomass derived materials. J Anal Appl Pyrolysis 62(2):331–349. https://doi.org/10.1016/S0165-2370(01)00129-2.
22. Garba A (2020) Biomass conversion technologies for bioenergy generation: an introduction. Biotechnological Applications of Biomass. IntechOpen.
23. Garcia CM, Teixeira S, Marciniuk LL, Schuchardt U (2008) Transesterification of soybean oil catalyzed by sulfated zirconia. Bioresour. Technol. 99(14):6608. https://doi.org/10.1016/j.biortech.2007.09.092.
24. Gerardi M H (2003) The microbiology of anaerobic digesters, John Wiley and Sons Inc., USA.
25. Giang TT, Lunprom S, Liao, Q, Reungsang A, Salakkam A (2019) Improvement of hydrogen production from Chlorella sp. Biomass by acid-thermal pretreatment. PeerJ 7:e6637. https://doi.org/10.7717/peerj.6637.
26. Goff MJ, Bauer NS, Lopes S, Sutterlin WR and Suppes GJ (2004) Acid-Catalyzed Alcoholysis of Soybean Oil. Journal of the American Oil Chemists' Society, 81:415–420.
27. Guran S (2020) Thermochemical conversion of biomass. Pract. Perspect. Sustain. Bioenergy A Syst. Think. Appr. 159–194. https://doi.org/10.1007/978-81-322-3965-9_8.
28. Gururani P, Bhatnagar P, Bisht B, Jaiswal KK, Kumar V, Kumar S, Rindin KG (2022) Recent advances and viability in sustainable thermochemical conversion of sludge to bio-fuel production, Fuel, 316:123351. https://doi.org/10.1016/j.fuel.2022.123351.
29. Hossain SMZ (2019) Biochemical conversion of microalgae biomass into biofuel. Chem Eng Technol 42:2594–2607. https://doi.org/10.1002/ceat.201800605.
30. Inayat A, Ahmed A, Tariq R, Waris A, Jamil F, Ahmed SF, Park YK (2022) Techno-economical evaluation of bio-oil production via biomass fast pyrolysis process: a review. Front Energy Res 9:770355. https://doi.org/10.3389/fenrg.2021.770355.
31. Jacobson K, Gopinath R, Meher L, Dalal A (2008) Solid acid catalyzed biodiesel production from waste cooking oil. Appl. Catal., B 85(1–2):86. https://doi.org/10.1016/j.apcatb.2008.07.005.
32. Jahirul MI, Rasul MG, Chowdhury AA, Ashwath N (2012) Biofuels production through biomass pyrolysis—a technological review. Energies (Basel) 5(12):4952–5001. https://doi.org/10.3390/en5124952.
33. Joshi NC, Sinha S, Bhatnagar P, Nath Y, Negi B, Kumar V, Gururani P (2024) A concise review on waste biomass valorization through thermochemical conversion, Current Research in Microbial Sciences, 6:100237. https://doi.org/10.1016/j.crmicr.2024.100237.
34. Kan T, Strezov V, Evans TJ (2016) Lignocellulosic biomass pyrolysis: a review of product properties and effects of pyrolysis parameters. Renew Sustain Energy Rev 57:1126–1140. https://doi.org/10.1016/j.rser.2015.12.185.
35. Khanal SK (2008) Anaerobic biotechnology for bioenergy production principles and applications, Wiley-Blackwell, Singapore.
36. Kumar A, Jones DD, Hanna MA (2009) Thermochemical biomass gasification: a review of the current status of the technology. Energies (Basel) 2(3):556–581. https://doi.org/10.3390/en20300556.
37. Kwon EE, Yi H, Jeon YJ (2013) Mechanistic investigation into water tolerance of non-catalytic biodiesel conversion. Appl Energy 112:388. https://doi.org/10.1016/j.apenergy.2013.06.038.

38. Laird DA, Brown RC, Amonette JE, Lehmann J (2009) Review of the pyrolysis platform for coproducing bio-oil and biochar. Biofuels Bioprod Biorefin 3(5):547–562. https://doi.org/10.1002/bbb.169.
39. Leung DYC, Guo Y (2006) Transesterification of neat and used frying oil: Optimization for biodiesel production. Fuel Process Technol 87(10):883. https://doi.org/10.1016/j.fuproc.2006.06.003.
40. Leung DYC, Wu X, Leung MKH (2010) A review on biodiesel production using catalyzed transesterification. Appl Energy 87(4):1083. https://doi.org/10.1016/j.apenergy.2009.10.006.
41. Li H, Niu S, Lu C, Li J (2016) Calcium oxide functionalized with strontium as heterogeneous transesterification catalyst for biodiesel production. Fuel 176:63. https://doi.org/10.1016/j.fuel.2016.02.067.
42. Solis JL, Alejo L, Kiros Y (2016) Calcium and tin oxides for heterogeneous transesterification of Babasssu oil (Attalea speciosa) J Environ Chem Eng 4(4):4870. https://doi.org/10.1016/j.jece.2016.04.006.
43. Li P, Miao X, Li R, Zhong J (2011) In Situ Biodiesel Production from Fast-Growing and High Oil Content Chlorella pyrenoidosa in Rice Straw Hydrolysate. J. Biomed. Biotechnol. 2011:141207. https://doi.org/10.1155/2011/141207.
44. Ma G, Hu W, Pei H, Jiang L, Ji Y, Mu R (2015) Study of KOH/Al2O3 as heterogeneous catalyst for biodiesel production via in situ transesterification from microalgae. Environ Technol 36(5):622. https://doi.org/10.1080/09593330.2014.954629.
45. Madhuvilakku R, Piraman S (2013) Biodiesel synthesis by TiO_2–ZnO mixed oxide nanocatalyst catalyzed palm oil transesterification process. Bioresour Technol 150:55. https://doi.org/10.1016/j.biortech.2013.09.087.
46. Matsumura Y (2015) Hydrothermal gasification of biomass. Recent Advances in Thermo-Chemical Conversion of Biomass. Elsevier, pp. 251–267. https://doi.org/10.1016/B978-0-444-63289-0.00009-0.
47. Melero JA, Iglesias J, Morales G (2009) Heterogeneous acid catalysts for biodiesel production: current status and future challenges. Green Chem. 11(9):1285. https://doi.org/10.1039/B902086A.
48. Mishra P, Krishnan S, Rana S, Singh L, Sakinah M, Ab Wahid Z (2019) Outlook of fermentative hydrogen production techniques: An overview of dark, photo and integrated dark-photo fermentative approach to biomass. Energy Strategy Rev 24:27–37. https://doi.org/10.1016/j.esr.2019.01.001.
49. Nagarajan D, Chang JS, Lee DJ (2020) Pretreatment of microalgal biomass for efficient biohydrogen production–Recent insights and future perspectives. Bioresour Technol 302:122871. https://doi.org/10.1016/j.biortech.2020.122871.
50. Nayono SE (2009) Anaerobic digestion of organic solid waste for energy production, KIT Scientific Publishing, Karlsruhe.
51. Nussbaumer T (2003) Combustion and co-combustion of biomass: fundamentals, technologies, and primary measures for emission reduction. Energy Fuels 17 (6):1510–1521. https://doi.org/10.1021/ef030031q.
52. Paul R, Silkina A, Melville L, Suhartin S, Sulu M (2023) Optimization of ultrasound pretreatment of microalgal biomass for effective biogas production through anaerobic digestion process. Energies 16:553. https://doi.org/10.3390/en16010553.
53. Perera SM, Wickramasinghe C, Samarasiri BKT, Narayana M (2021) Modeling of thermochemical conversion of waste biomass–a comprehensive review. Biofuel Res J 8 (4):1481–1528. https://doi.org/10.18331/BRJ2021.8.4.3.
54. Phanduang O, Lunprom S, Salakkam A, Liao Q, Reungsang A (2019) Improvement in energy recovery from Chlorella sp. Biomass by integrated dark-photo biohydrogen production and dark

fermentation-anaerobic digestion processes. Int J Hydrog Energy 44:23899–23911. https://doi.org/10.1016/j.ijhydene.2019.07.103.
55. Pourzolfaghar H, Abnisa F, Daud WMAW, Aroua MK (2016) A review of the enzymatic hydroesterification process for biodiesel production. Renewable Sustainable Energy Rev. 2016, 61, 245. https://doi.org/10.1016/j.rser.2016.03.048.
56. Puhan S, Vedaraman, Ram BVB, Sankarnarayanan G and Jeychandran K (2005) Mahua Oil (Madhuca indica Seed Oil) Methyl Ester as Biodiesel-Preparation and Emission Characterstics. Biomass and Bioenergy, 28:87–93. https://doi.org/10.1016/j.biombioe.2004.06.002.
57. Ram M, Mondal MK (2022) Biomass gasification: a step toward cleaner fuel and chemicals. Biofuels and Bioenergy. Elsevier, 253–276.
58. Rashid U, Anwar F (2008) Production of biodiesel through optimized alkaline-catalyzed transesterification of rapeseed oil. Fuel 87(3):265. https://doi.org/10.1016/j.fuel.2007.05.003.
59. Rashid U, Anwar F, Moser BR, Ashraf S (2008) Production of sunflower oil methyl esters by optimized alkali-catalyzed methanolysis. Biomass Bioeng 32(12):1202. https://doi.org/10.1016/j.biombioe.2008.03.001.
60. Ravichandran SR, Venkatachalam CD, Sengottian M, Sekar S, Kandasamy S, Prasath K, et al. (2022) A review on hydrothermal liquefaction of algal biomass on process parameters, purification and applications. Fuel 313:122679. https://doi.org/10.1016/j.fuel.2021.122679.
61. Razzak SA, Hossain MM, Lucky RA, Bassi AS, Lasa H (2013) Integrated CO_2 capture, wastewater treatment and biofuel production by microalgae culturing—a review. Renewable Sustainable Energy Rev 27:622. https://doi.org/10.1016/j.rser.2013.05.063.
62. Ren J, Cao JP, Zhao XY, Yang FL, Wei XY (2019) Recent advances in syngas production from biomass catalytic gasification: a critical review on reactors, catalysts, catalytic mechanisms and mathematical models. Renewable Sustainable Energy Rev 116:109426. https://doi.org/10.1016/j.rser.2019.109426.
63. Ren XY, Ghazani MS, Zhu H, Ao WY, Zhang H, Moreside E, et al. (2022) Challenges and opportunities in microwave-assisted catalytic pyrolysis of biomass: a review. Appl Energy 315:118970. https://doi.org/10.1016/j.apenergy.2022.118970.
64. Riva L, Wang L, Ravenni G, Bartocci P, Buø TV, Skreiberg Ø, Nielsen HK (2021) Considerations on factors affecting biochar densification behavior based on a multiparameter model. Energy 221:119893. https://doi.org/10.1016/j.energy.2021.119893.
65. Saravanakumar A, Haridasan TM, Reed TB (2010) Flaming pyrolysis model of the fixed bed cross draft long-stick wood gasifier. Fuel Process Technol 91(6):669–675. https://doi.org/10.1016/j.fuproc.2010.01.016.
66. Schwab AW, Bagby MO, Freedman B (1987) Preparation and properties of diesel fuels from vegetable oils. Fuel 66 (10):1372. https://doi.org/10.1016/0016-2361(87)90184-0.
67. Shuit SH, Hoong S, Lee KT, Kamaruddin AH, Yusup S (2010) Reactive extraction and in situ esterification of Jatropha curcas L. seeds for the production of biodiesel. Fuel 89 (2):527. https://doi.org/10.1016/j.fuel.2009.07.011.
68. Sikarwar VS, Zhao M, Clough P, Yao J, Zhong X, Memon MZ, Fennell PS (2016) An overview of advances in biomass gasification. Energy Environ Sci 9(10):2939–2977. https://doi.org/10.1039/C6EE00935B.
69. Singh SP, Singh D (2010) Biodiesel production through the use of different sources and characterization of oils and their esters as the substitute of diesel: A review. Renewable Sustainable Energy Rev 14 (1):200. https://doi.org/10.1016/j.rser.2009.07.017.
70. Sivabalan K, Hassan S, Ya H, Pasupuleti J (2021) A review on the characteristic of biomass and classification of bioenergy through direct combustion and gasification as an alternative power supply. In: Journal of Physics: Conference Series, 1831. IOP Publishing, 012033.

71. Sun H, Hu K, Lou H, Zheng X (2008) Biodiesel Production from Transesterification of Rapeseed Oil Using KF/Eu_2O_3 as a Catalyst. Energy Fuels 22(4):2756. https://doi.org/10.1021/ef700778r.
72. Sung YJ, Yu BS, Yang HE, Kim DH, Lee JY, Sim SJ (2022) Microalgae-derived hydrogen production towards low carbon emissions via large-scale outdoor systems. Bioresour Technol 364:128134. https://doi.org/10.1016/j.biortech.2022.128134.
73. Tubino M, Junior JGR, Bauerfeldt GF (2016) Biodiesel synthesis: A study of the triglyceride methanolysis reaction with alkaline catalysts. Catal. Commun. 75:6. https://doi.org/10.1016/j.catcom.2015.10.033.
74. Tumuluru JS, Ghiasi B, Soelberg NR, Sokhansanj S (2021) Biomass torrefaction process, product properties, reactor types, and moving bed reactor design concepts. Front Energy Res 9:728140. https://doi.org/10.3389/fenrg.2021.728140.
75. Wang G, Dai Y, Yang H, Xiong Q, Wang K, Zhou J, Wang S (2020) A review of recent advances in biomass pyrolysis. Energy Fuels 34(12):15557–15578. https://doi.org/10.1021/acs.energyfuels.0c03107.
76. Wang YZ, Li YJ (2023) Sorption-enhanced steam gasification of biomass for H2-rich gas production and in-situ CO_2 capture by CaO-based sorbents: a critical review. Applications in Energy and Combustion Sci 14:100124. https://doi.org/10.1016/j.jaecs.2023.100124.
77. Xu YH, Li MF (2021) Hydrothermal liquefaction of lignocellulose for value-added products: mechanism, parameter and production application. Bioresour Technol 342:126035. https://doi.org/10.1016/j.biortech.2021.126035.
78. Xue Y, Chen HY, Zhao WN, Yang C, Ma P, Han S (2016) A review on the operating conditions of producing bio-oil from hydrothermal liquefaction of biomass. Int J Energy Res 40:865–77. https://doi.org/10.1002/er.3473.
79. Yu H, Liu Y, Dong W, Li W, Li R (2007) Experimental study on the biomass flash pyrolysis. In: Challenges of Power Engineering and Environment: Proceedings of the International Conference on Power Engineering 2007. Springer, Berlin Heidelberg, 1152–1154.
80. Zhang J, Zhang X (2019) The thermochemical conversion of biomass into biofuels. Biomass, Biopolymer-Based materials, and Bioenergy. Woodhead Publishing, 327–368.
81. Zhang Y, Cheng J, He Y, Yuan J (2022) Photo-fermentative hydrogen production performance of a newly isolated Rubrivivax gelatinosus YP03 strain with acid tolerance. Int J Hydrog Energy 47:20784–20792. https://doi.org/10.1016/j.ijhydene.2022.04.198.
82. Zhang Y, Wan L, Guan J, Xiong QA, Zhang, Jin X (2020) A review on biomass gasification: effect of main parameters on char generation and reaction. Energy Fuels 34 (11):13438–13455. https://doi.org/10.1021/acs.energyfuels.0c02900.

Biodiesel and Bioethanol Production

4

Abstract

Concerns about the depletion of petroleum supplies, energy security, air pollution, and global warming have fueled global interest in finding sustainable or renewable alternative energy sources. Many countries believe that biofuels will soon be essential for self-sufficient energy generation as well as reducing emissions of air pollutants and greenhouse gases such carbon dioxide, nitrogen oxide, sulfur oxides, and methane. Biofuels are an eco-friendly and sustainable renewable energy source that are created from biomass such as agricultural produce or organic waste materials. Microalgae often degrade or ferment vegetable and animal organic materials to generate liquid or gas fuels such as methane, ethanol, and hydrogen through thermal and chemical processes. Biofuels include bioethanol, biodiesel, biomethane, and biobutanol. First-generation biofuels employed sugars found in edible crops, but second-generation biofuels may use lignin, albeit it is extremely refractory, and cellulose from sources such as wood pulp. The conversion of non-biodegradable organic molecules to fermentable sugars necessitates extra technological procedures such as thermal, chemical, and enzymatic treatments, raising the cost. Microalgae play a vital role in the natural carbon cycle, and they may efficiently use carbon resources in water and soil. The ability of microalgae to reduce CO_2 emissions and generate oil with high productivity has been demonstrated. Renewable biofuels are required to replace petroleum-derived transportation fuels, which contribute to global warming and are limited in availability. Biodiesel and bioethanol are two possible renewable fuels that have received the greatest attention. As illustrated here, biodiesel and bioethanol produced from agricultural crops using current technologies cannot replace fossil-based transportation fuels indefinitely, but there is a solution. Biodiesel from microalgae appears to be the only renewable biofuel capable of totally replacing petroleum-derived transportation fuels without jeopardizing food and agriculture supplies. Most prolific oil crops, such as oil palm, fall well

H. N. Akolkar et al., *Bioenergy from Microalgal Biomass*, Synthesis Lectures on Renewable Energy Technologies, https://doi.org/10.1007/978-3-032-03434-2_4

short of microalgae in terms of producing the requisite volumes of biodiesel on a sustainable basis (Brennan and Owende in Renewable and sustainable energy reviews 14:557–577, 2010 [3]; Chisti in Biotechnology advances 25:294–306, 2007 [8]; Mata et al. in Renewable and sustainable energy reviews 14:217–232, 2010 [30]). Algae-based biofuels have recently been created and categorized as third-generation biofuels. Because microalgae have no lignin, they may be easily converted to monosaccharides for ethanol generation. Furthermore, microalgae grow quickly, produce biofuel efficiently, and have a short harvesting cycle, resulting in cheaper costs than other feedstocks. Algal oil and biomass are easily turned to diesel and gasoline. Microalgae may be grown anywhere, including wastelands, coasts, and oceans, as long as photosynthesis is feasible. Furthermore, they don't compete with culinary crops for cultivation area or space.

Keywords

Microalgae • Biomass • Harvesting • Biodiesel • Bioethanol • Renewable • Biofuels • Ferment • Photosynthesis

4.1 Microalgal Biodiesel Extraction and Processing

The recovery of crude oil from microalgal biomass and its conversion to biodiesel are unaffected by whether the biomass is generated in raceways or photobioreactors. As a result, the cost of creating biomass is the sole important consideration in comparing photobioreactors with raceways for microalgal biodiesel production (Fig. 4.1).

The cost of creating microalgal biodiesel may be significantly decreased by employing a biorefinery-based production plan, enhancing microalgae capabilities through genetic engineering, and making breakthroughs in photobioreactor design.

Depending on the microalgae employed, it may be possible to extract additional high-value compounds. Several research have improved the area of biofuel synthesis

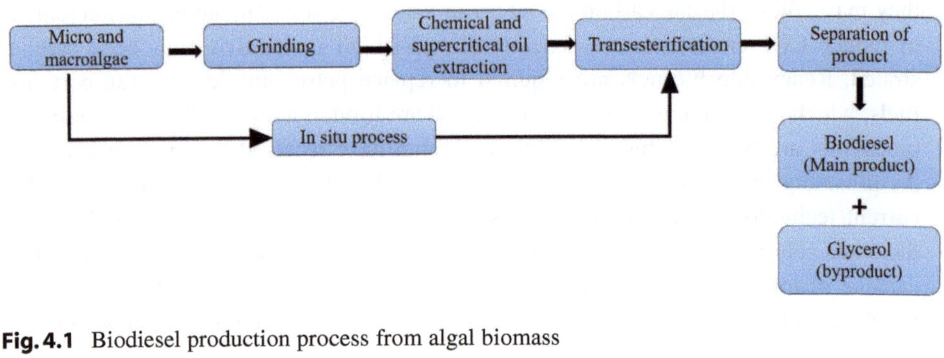

Fig. 4.1 Biodiesel production process from algal biomass

from microalgae biomass, including extensive and evaluative investigations. Kröger and Müller-Langer [22] evaluated the essential aspects of algal biofuel production methods (transesterification and hydro-processing) and their related economic and environmental problems. Alternatively, Mubarak et al. [34] examined several techniques of extracting lipids from microalgae biomass for biodiesel generation. The study found that ultrasonication and microwave-assisted approaches in solvent extraction improve microalgae lipid output. This extensive analysis focused on the utilization of microalgae for biodiesel production. Qari et al. [48] emphasized the value of microalgae as a renewable energy source, as well as the limitations that come with it.

Furthermore, Osman et al. [40] examine developments in biomass conversion to biofuels and their environmental effect using life cycle assessment. Zanuso et al. [73] also explore the benefits and drawbacks of immobilizing hydrolytic lignocellulosic biomass enzymes and the hydrolysis of various lignocellulosic biomasses, whilst Srivastava et al. [64] look at bio-wastes and the employment of microbial systems to break them down. Moving further, Raheem et al. [49] recently conducted a study of several conversion techniques used in syngas and biodiesel production, including pyrolysis, gasification, hydrothermal liquefaction, fermentation, and transesterification.

4.1.1 Lipid Extraction Processes

Microalgae's high lipid content makes them potential for biofuel generation [19]. This is because a variety of parameters, including species type, beginning lipid content, and amount of biomass treated per unit time, impact the selection of an extraction technique [52]. The following lipid extraction procedures are commonly mentioned in the literature: Bligh and Dryer, microwave oven, supercritical fluid extraction, pressurized fluid extraction, ultrasound, sonication, and soxhlet extraction.

4.1.1.1 Solvent Extraction

Phukan et al. [46] evaluated the utilization of thirteen different solvents for lipid extraction from *Chlorella* sp. This was because this species generates more oil than the majority of other algae species. Excess lipids (>10%) were removed using chloroform, ethanol, and hexane solvents for an optimal period of 3 h. This might be attributed to the potential development of volatile degradation products. Furthermore, the binary combination of 1:1 chloroform: ethanol demonstrated higher efficiency with a lipid production amount of 11.76%, whereas the best single solvent, chloroform, produced 10.78%.

The binary mixtures high efficiency can be ascribed to the fact that using solvents other than binary combination would be impossible for upscaling and could potentially raise biodiesel production expenses. The chloroform generated more than 75% fatty acid methyl esters (FAME), which was 58% and 25% higher than that of hexane and ethanol, respectively, which is also depends on polarity and solubility of chloroform [66]. From these studies, for lipid extraction ethanol, chloroform, and hexane are characteristically more effective solvents.

The efficiency of the extraction process is better established for production yields and economics. Although Ramaluckan et al. [51] noted that using solvents other than binary combinations might raise biodiesel production costs. Naghdi et al. [36] investigated the use of a chloroform, methanol and hexane solvent mixture in extracting *Ulva fasciata* algae oil. The study found that a 1:1:2 volume ratio of these solvents had the highest oil yield of 9.85%. In addition, the acid and saponification values of the algal oil were 13.73 mg and 194.7 mg KOH/g, respectively, showing a saturated fatty acid content of 75.43%. There is also existence of 12-hydroxyl-9-octadecenoic acid required for biodiesel generation. In spite of the evident appropriateness of *U. fasciata* for biodiesel synthesis, this method is not cost viable. Ramaluckan et al. [51] and Patil et al. [44] studied the use of leftover biomass for bioethanol production. They also investigated the effectiveness of pretreatments in improving lipid recovery from algal slurry.

Study shows lipid recovery from wet *C. muelleri* algal biomass by processing with UV radiation and a hexane: ethanol solvent combination due to which there is no need to dry the algal biomass. By using co-solvent combination for pretreatments, availability of intracellular lipids for recovery can be increased. This approach may lead to the development of a more cost-effective microalgae oil manufacturing pathway. This is because the remaining biomass was studied and discovered to have carbs for the fermentation process of bioethanol generation. Similarly, adding hexane and ethanol to UV-C and thermally processed algal slurries resulted in a more than doubling of total extractable and transesterifiable lipids.

More recent research has been spurred by the possibility that this process is not economically driven. Yang et al. [72] investigated supercritical carbon dioxide extraction of lipids from algal biomass utilizing azeotropic co-solvents hexane and ethanol. This study found that supercritical carbon dioxide extraction using azeotropic co-solvents is an excellent method for obtaining total lipids/biooils from algae. This is due to numerous factors. Initially, 20–32% of the important eicosapentanoic acid was recovered from the algae. Second, the technique used has the following advantages: higher total lipid yield and selectivity, better lipid purity, and shorter extraction time while preserving thermolabile molecules. Third, the heating value for neutral algal lipid was 40.36 MJ/kg, which was comparable to petroleum-derived crude. Similarly, a maximum total algal lipid yield of 31.37% was obtained under reaction parameters of 340 bar pressure, (12 hexane + 1 ethanol) to algae/solid ratio, 80 °C, and 60 min reaction temperature and duration, respectively, under CO_2 flowrate regulated circumstances. The capacity of the cosolvent to boost overall lipid yield impacted the decision to employ a large quantity. So, these reaction conditions have a considerable impact on lipid production, long chain fatty acid recovery, and extraction process economics. Future research should focus on clarifying how the procedure might be improved to increase the lipid's heating value.

4.1.1.2 The Use of Deep Eutectic Solvents

Cheng et al. [7] evaluated the effect of aqueous deep eutectic solvents (DES) on lipid extraction from *Chlorella* sp. They used a blend of ethyl acetate and ethanol solvents to improve lipid recovery. The effect on the pretreatment of *chlorella* sp. biomass of three aqueous DES (1. aqueous urea-acetamide (aUA); 2. aqueous choline-chloride (a Ch-O); 3. ethylene glycol (a Ch-EG)) was calculated by measuring lipid content, lipid recovery efficiency, along with carbohydrate content. Pretreatment with aqueous DES substantially improved lipid recovery ($P < 0.05$), demonstrating a DES-dependent effect. Furthermore, treatment of the algae with the DES (a Ch-O, a Ch-EG, and aU-A) decreased total carbohydrate content recovery in algal biomass by 24.21, 20.14, and 27.89%, correspondingly. This is due to cytoplasm reduction in microalgae cells [43]. Thus, lipid recovery using a combination of ethyl acetate and ethanol enhanced following aqueous DES pretreatment, signifying that aqueous DES can increase the efficiency of lipid extraction from *Chlorella* sp. biomass. This is due to their capacity to disintegrate the cell wall. Similarly, Piligaev et al. [47] explored extraction yield was enhanced by acid DES cell disruption. The effects of solvent quantity, DES, biomass mass ratio, extraction duration, and temperature on FAME were explored while comparing one-step along with two-step biodiesel synthesis procedures with help of microalgae biomass. DES extracts FAME can result in microalgae-based biodiesel in a single step. First, the total 30% higher FAME content extracted utilizing the one-step approach with DES treatment compared to the two-step method. Similarly, the two-step technique increased FAME content when compared to the lack of DES pretreatment. Next, a rise in DES led in an increase in cell and FAME destruction when using the one-step technique. Because microalgae cell death is proportional to DES strength, FAME recovery may be improved accordingly. This explains why the study found that 1:3 mol/mol of chlorine-acetic acid was more successful than other DES ratios in extracting FAME from *Chlorella* sp. and *Chlororococcum* sp. biomass.

4.1.2 Biorefinery Based Production

A biorefinery, like a petroleum refinery, converts every component of biomass raw material into usable output.

Because all biomass components are used, the overall cost of manufacturing any particular product is reduced. Integrated biorefineries are already in operation in Canada, the United States, and Germany, generating biofuels and other products from crops including maize and soybeans. This technology can be used to lower the cost of producing microalgal biodiesel. In addition to oils, microalgal biomass includes considerable amounts of proteins, carbs, and other nutrients [59]. As a result, the remaining biomass from biodiesel manufacturing methods might possibly be utilized as animal feed. Some of the remaining biomass might be utilized to make methane through anaerobic digestion, which would generate the electrical power required to run the microalgal biomass production plant

[31, 55]. Excess electricity might be sold to help cover the costs of creating biodiesel. Although using microalgal biomass directly to create methane via anaerobic digestion is theoretically viable, it cannot compete with the several alternative low-cost organic substrates available for anaerobic digestion. Nonetheless, algal biomass waste left behind after oil extraction has the potential to be converted into methane. A microalgal biorefinery is capable of producing biodiesel, animal feed, biogas, and electricity all at the same time.

4.2 Transesterification Reaction and Catalysts

Currently, research is looking at the use of biowaste-derived catalysts for more sustainable biodiesel synthesis. As a result, Karpagam et al. [20] recently provided more insight into the activity of waste egg shell-derived Nano-CaO catalyst on transesterification and its optimization by varying Nano-CaO catalyst dosage, time, and temperature using a five-level central composite design based full factorial response surface model. The greatest fatty acid methyl ester (FAME) yield was 93.44% at ideal circumstances of 2.06% (wt/wt) catalyst, 180 min, 60 °C, and a constant string speed of 200 rpm. Furthermore, the catalyst was reusable and stable for up to six cycles, confirming an average biodiesel output of 85.22%. This is significant because catalyst recycling eliminates the issues of solvent use, energy expenditure, and operational constraints. This is why heterogeneous catalysts are chosen over homogeneous catalysts. Taken together, our findings demonstrate that the Nano-CaO catalyst generated from discarded egg shell is both sustainable and cost-effective for biodiesel generation. Similarly, a recent work [37] produced and evaluated bio-nanoparticles from discarded seashells as catalysts for the direct transesterification of M60 from *Coelastrella* spp. to FAME conversion. The bio-nano $CaCO_3$ and CaO provided the highest FAME proportions (20.6 ± 1.7 and 20.3 ± 1.1%, respectively). In comparison to traditional acid catalysis, bio-nanoparticle mediated transesterification resulted in a 1.29 times increase in FAME. The basic sites found in bio-nanoparticle catalysts are the active sites responsible for FAME conversion, and they are both easily accessible and densely populated. Bio-nanoparticle-mediated transesterification is found superior to traditional acid catalysis. Bio-waste-derived catalysts provide a more environmentally friendly and cost-effective method of producing biodiesel. Other recent studies by Das et al. [9], Wadood et al. [69], and Cheirsilp and Louhasakul [4] support the utilization of bio-wastes-derived catalysts for better biodiesel production.

4.2.1 Single-Step Methods

Patil et al. [45] investigated direct transesterification of microbial lipids into biodiesel. This lowered the raw material and processing costs of biodiesel manufacturing. Transesterification single step technique produced a biodiesel yield more than 70% in one hour. But it skipped the cell drying and lipid extraction processes. Direct transesterification process has the potential to the industrialization of significant oil production from microorganisms and industrial waste. The direct conversion of algal biomass into biodiesel was revealed by Rathnam et al. [53] using supercritical methanol (SCM) and microwave-assisted (MW) transesterification techniques. The MW technique boosted algal extraction efficiency by 60%, shortened extractive transesterification time to 4–5 min, and increased yield by 80.13% based on total lipid content.

The non-catalytic SCM generated clean solvents and catalyst residues while consuming less energy throughout the separation and purification processes. This decrease in energy usage is attributed to the simpler purification phase. This is because the study found that the GCMS-revealed FAME percent purity may have affected the effective separation and purification of the product during the supercritical process. Also, sample weight loss due to thermal degradation (from 10, 50, and 90% of the initial weight) was found at temperatures of 195, 362, and 430 °C, corresponding to the increasing deterioration of the weight losses. Overall, these data show why single-step approaches are considered a manner of decreasing chemical and energy use in the biodiesel synthesis process.

4.2.1.1 Supercritical Transesterification (SCT)

The employment of catalysts in microalgae biodiesel conversion is an old traditional approach that can result in soap production when the free fatty acid level exceeds 0.5%, as well as lower biodiesel output when water is present in the feedstock. This is why SCT incorporating catalyst absence is now being studied. Aside from the fact that the absence of catalysts may result in delayed reaction processes, their use may entail additional expenses to the process [70]. As a result, Chauhan et al. assessed the feedstock quality for the direct conversion of Chlorella sp. FC2 IITG microalga into biodiesel using supercritical methanol transesterification (SCMT). The greatest FAME yield was 96.9% at optimal values of 52% (wt/wt) lipid content, 5.75 mL/g water content, and 115 mL/g methanol loading.

Furthermore, Knothe [21] investigated the impact of process parameters in dry algae single-step conversion to ethanol (in wt/vol) ratios of 1:6–1:15, reaction temperature of 245–270 °C, and reaction duration of 2–30 min on FAEE yield under supercritical ethanol conditions. According to the study, this "green" conversion method has the potential to provide an energy-efficient and cost-effective approach for producing sustainable biodiesels. The calorific value of FAEEs was 43 ± 1.8 MJ/kg, similar to ordinary diesel fuel. This calorific value exceeds that of FAME generated from the same feedstock.

More recently, the conversion of microalgae oil to FAME by SCM transesterification was optimized using sequential hybrid optimization with RSM, ANN, and genetic algorithm (GA) [63]. The work aimed to improve the non-catalytic transesterification of microalgae oil to biodiesel using supercritical methanol conductions. GA created an optimization condition for SCM transesterification with a temperature of 285.2 °C, a time of 26.5 min, and a MeOH: oil molar ratio that yielded 99.16% conversion efficiency. The SCM transesterification yield was around 99–100% conversion efficient, making the method a zero waste generating procedure for biodiesel synthesis.

Anyway, Aghilinategh et al. [1] reported on their study on altering the supercritical transesterification process. They carried out a one-pot biodiesel synthesis from Chlorella vulgaris in SCM using TiO_2 and $SrTiO_3$ nanocatalysts and co-solvents. The modification procedure demonstrates that preparing catalysts via a photochemical technique prevents leaching even under strong reaction conditions. This is because the use of n-hexane co-solvent resulted in the maximum yields of biodiesel and other products due to the material's extremely low dielectric constant. This is owing to its capacity to increase the rate of mass transfer caused by improved biomass breakdown.

4.2.1.2 Microwave-Assisted Transesterification

The application of MW technology greatly decreases reaction time, enhances biodiesel output, and simplifies catalyst recovery operations. Accordingly, Bhuana et al. [2] investigated biodiesel generation using in situ transesterification of Chlorella sp. This was accomplished by testing the effects of acid catalyst concentration, microwave power, and retention time. This study demonstrated that the Box-Behnken Design optimization technique allows for a high biodiesel output of appropriate quality. This is because 63.6% biodiesel production was achieved using 370 W microwave power, 0.2 M catalyst concentration, and an 82.7-min transesterification duration. Moving ahead, Lawan et al. [24] demonstrated the role of a microwave reactor system in efficient and sustainable waste lard biodiesel manufacturing.

Continuing the work of Bhuana et al. [2], Teo and Idris [65] described the utilization of MW for increased extractive-tansesterification of algae lipids from dry *Chlorella* sp. MW irradiation can be used to extractive-transesterify dry algal biomass instead of the two-step extraction and transesterification procedure that occurs simultaneously. This is because the MW helps to increase algal lipid and fatty acid ethyl ester (FAEE) production. This is proved by the FAEE conversion of algal lipids, which were 96.2%, 94.3%, and 78.1% for MW, MW with hexane, and traditional Bligh and Dyer (BD), respectively. These FAEE conversions were calculated using maximal lipid yields of 20.1, 20.1, and 13.9%. Also, the hexane solvent lowered the severity of the reaction conditions, resulting in equivalent lipid and FAEE yields.

This is because hexane and ethanol co-solvents can boost the extractive ability of FFA, resulting in higher transesterification reaction yields. Similarly, Rathnam et al. [53] studied the effect of microwave irradiation on the simultaneous extraction and transesterification of dry algal biomass into diesel. The goal of this work was to optimize microwave transesterification using RSM. In-situ transesterification proven to be a quick and straightforward way to produce biodiesel from dry biomass. Methanol speeds up the reaction because it is a powerful microwave absorption material, with the OH- group radical dominating the microwave spectrum and causing localized superheating.

4.2.2 Transesterification Catalysts

Transesterification catalysts are broadly categorized into two types: homogeneous and heterogeneous. Catalysts are classified into several types, the most common of which are acidic and basic. Biocatalysts were recently included as a third categorization category (Fig. 4.2).

4.2.2.1 Homogeneous Transesterification Catalysts

Using homogeneous catalysts has created a number of challenges. Because they are in the same liquid phase as the reaction mixture, the separation method might be expensive. Furthermore, soap and water production necessitates more homogeneous catalysts, and the synthesis of byproducts from high FFA biomasses in homogeneous catalysis lowers

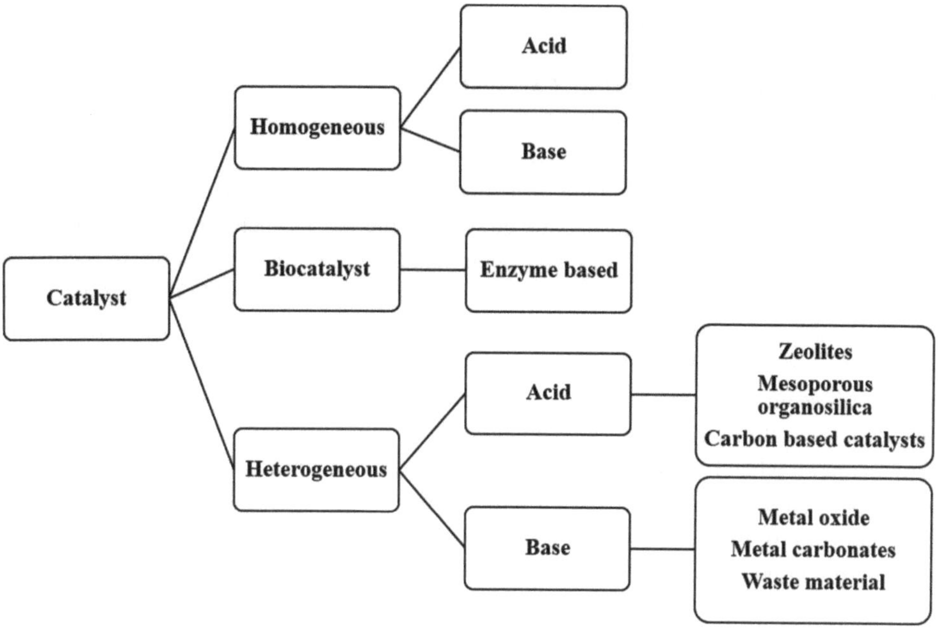

Fig. 4.2 Algal transesterification catalysts classification

biodiesel yield [50]. As more research has been undertaken using base catalysts for the transesterification of algal lipids to biodiesel, it is probable that alkaline catalysts have been more active, resulting in higher conversion rates and so being more efficient in this process. Some experiments used homogeneous base catalysts for algal lipid transesterification. NaOH and KOH are among the most inexpensive and widely available catalysts, hence they are often used for transesterification of several lipid classes [18]. However, difficulties in commercializing these catalysts, such as reaction system corrosion, waste production, and limited recyclability, led researchers to develop solid acid catalysts.

4.2.2.2 Heterogeneous Transesterification Catalysts

Heterogeneous catalysts have a separate phase that differs from the reaction mixture, making separation and purification easier. Furthermore, they prevent equipment corrosion and make recycling easier [50]. In transesterification using heterogeneous basic catalysts, the strength and density of basic sites have a considerable impact on reaction rate, unlike Lewis or Brønsted basicity parameters. Many biomass-derived lipids have been transesterified using basic polymers, oxides (CaO, MgO, SrO, and BaO) and carbonates ($CaCO_3$, $MgCO_3$, and $Ca(OCH_3)_2$), basic zeolites, and first-row transition metal oxides. Furthermore, the sulphonic acid groups on the Nafion NR50 surface improved the catalyst's activity in palmitic acid esterification, triglyceride and diglyceride transesterification, and phosphocholine synthesis. Sorbitan monopalmitate had a very low conversion rate in the presence of all catalysts due to an ester bond between fatty acid and sugar molecules that prevented fatty acid from reacting.

4.2.2.3 Biotransesterification Catalysts

Enzymatic transesterification of lipids has lately grown in favor due to its gentle reaction conditions and lack of soap or byproduct generation. Biotransesterification catalysts for algae-to-biodiesel conversion are classified as free, immobilized, cell-bound, or extracellular enzymes. However, because the enzyme is the most expensive component of this unique catalytic technology, efforts have been made to immobilize it in order to reduce the cost of the process while also enhancing the catalyst's recyclability [35, 39, 57, 60]. Lipase immobilization may be performed using a variety of ways, one of the most common being covalent bonding, which results in enhanced heat stability and enzyme interaction [34]. Polar lipids seem to deactivate the immobilized Rhizopus oryzae lipase more quickly. Furthermore, the stability (three cycles) and activity of non-immobilized whole-cell catalysts were increased when hexane-extracted neutral saponifiable lipids were used as the reactant instead of ethanol-hexane extracted ones due to the presence of less polar lipids [28]. Lee et al. [25] improved the activity of the Novozyme 435 (immobilized lipase) biocatalyst by using dimethyl carbonate (DMC) as the reaction medium and acyl acceptor.

4.3 Bioethanol Production Through Fermentation

Industrial implementation of microalgae biofuel technology has always been hampered by its economic feasibility. To improve the feasibility of bioethanol production from microalgae, Chlorella vulgaris FSP-E biomass was fermented under high-solid-loading conditions. First, Chlorella vulgaris FSP-E was grown to yield carbohydrate-rich microalgae biomass. Then, several ethanol-producing bacteria were tested. Saccharomyces cerevisiae FAY-1 showed no inhibition when fermenting high starting glucose concentrations and was used for the fermentation studies with high solids loading. The acid hydrolysis process was also optimized at large biomass loading. The fermentation of microalgal biomass hydrolysate resulted in a higher final ethanol concentration and yield. Fermenting microalgae biomass can highly enhance microalgal bioethanol production (Fig. 4.3).

4.3.1 Microalgae Biomass Pretreatment

In all types of biorefineries, a critical step is raw material of aquatic biomass for production of bioenergy. After pretreatment intracellular compounds such as carbohydrates, proteins, and lipids, as well as added-value compounds such as short chain oligosaccharides, antioxidants, pigments become available for further processes [11, 14]. The primary pretreatment of microalgae biomass leads to breakdown of the cell wall and the alteration of the structure of intracellular carbohydrates. Carbohydrate is found in cell wall as cellulose and in the plastids as starch, which serves as the primary sugar reserve. The biomass with the maximum sugar content can be used for bioethanol production [5]. The cell wall

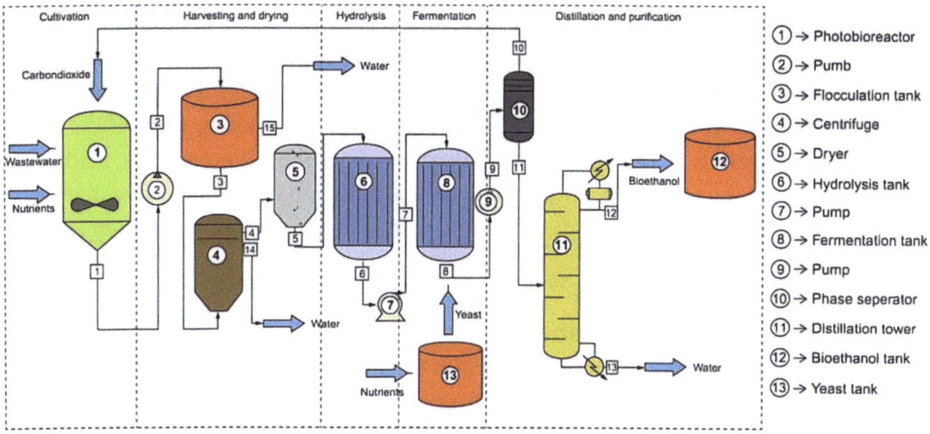

Fig. 4.3 Flow diagram of bioethanol production from microalgae [41]

is made up of cellulose, pectin (polygalacturonic acid) and sulfated polysaccharides that can be infused with inorganic elements including calcium carbonate, silica, and magnesium [5, 56]. Exterior cell wall is primarily composed of polysaccharides such as pectin, agar, alginate, and algaenan polymers, whereas interior cell wall is composed of pectin, fucans, hemicellulose, and glycoproteins in a cellulose, microfibril matrix with traces of fucose, xylose, rhamnose, arabinose, and galactose [62]. Depending on the composition, the cell wall of *Chlorella* spp. is more solid and hard making it more difficult to break but in *Scenedesmus* sp. and the cyanobacteria *Arthrospira* sp. cell wall is soft. The semi-crystalline granulose of starch made from amylose polymers and highly branched amylopectin [10, 17]. Starch granules in their crystalline form contain less water, which increases their stability and makes them more difficult to hydrolyze by enzymes. As a result, they must be gelatinized using chemical, physicochemical, or biological pretreatment procedures, or a combination of these [6, 7, 15, 23]. These include acid, alkaline, or biological pretreatments, as well as other approaches such as ultrasound, microwave, electric pulses, and so on [32, 74].

4.3.1.1 Enzymatic Pretreatment

Enzymatic hydrolysis has proven to be more effective than other pretreatment procedures [11, 27]. Enzymatic hydrolysis has several advantages, including excellent specificity, little need for harsh conditions, and ease of implementation on an industrial scale. Enzymatic hydrolysis uses a variety of enzymes, including cellulases, amylases, and amyloglucosidases, which are designed to hydrolyze cell wall polysaccharides. It is crucial to note that depending on the kind of microalgae, the cell wall may or may not be permeable to other chemicals, and specialized enzymes are seldom required to break the cell wall since these enzymes may reach internal carbohydrates.

4.3.1.2 Hydrothermal Pretreatment

Hydrothermal processing is an alternate method for breaking down microalgae cell walls and gelatinizing the intracellular starch. This process is also used to pre-treat agro-industrial wastes (such as wheat straw, sugarcane bagasse, and softwoods) as well as aquatic biomass (seaweed). This approach uses temperatures ranging from 60 to 180 °C and response durations of less than 60 min. Furthermore, in this pretreatment, acids, alkalis, or merely water can be utilized as reaction catalysts [5, 58]. The acid-based approach typically uses 1–10% concentrations of sulfuric acid and hydrochloric acid at temperatures ranging from 60 to 180 °C.

4.3.1.3 Ultrasound Pretreatment

Ultrasound technology is frequently utilized in organic chemistry to accelerate chemical processes and extract bioactive chemicals from a variety of plant species. The ultrasonic process includes the employment of sound waves that pass through a liquid medium and generate zones of compression and rarefaction. In this pressure shifts occur, resulting in the cavitation phenomena which creates bubbles in the elastic medium. By using this technology, we can break the cell wall of microalgae. This is achieved as bubbles burst on a cell wall, the pressure and high temperature generate microjets. This process allows the solvent to permeate the raw material by tearing the cell wall [29].

4.3.1.4 Microwave Pretreatment

Velazquez-Lucio et al. [67] pretreated lignocellulosic residues to produce second generation bioethanol and high value compounds. This alternative energy works by interacting with the solvents polar water molecules. This produces hot nuclei allowing for extremely efficient and quick heating. This heating is accomplished through two mechanisms. First, the rotation of the dipoles, in which the polar molecules attempt to align in the rapidly changing electromagnetic field produced by microwaves. This is followed by ionic conduction, which consists of the instant superheating of the ionic substance due to the friction of the ionic molecules caused by the movement that generates the electric field [61].

4.3.1.5 Pulse Electric Field Pretreatment (PEF)

The PEF pretreatment method is a simple electricity-based procedure that is non-thermal and executed in a short period of time (nanoseconds to milliseconds) but with a large amplitude (100–300 V cm^{-1} to 300 kV cm^{-1}). This pretreatment is effective at both low and high cell concentrations in liquid, and hence may be employed directly with the microalgal culture. This therapy produces a phenomenon known as electroporation or electropermeabilization. In reality, the electric field generated causes a power differential across the cellular membrane, and electroporation occurs when a particular threshold value is surpassed (0.5–1.5 V) [13, 68].

4.3.1.6 Mechanical Methods

Conventional mechanical methods are the most widely employed procedures in the business, providing positive results on a large scale as well as great recovery of the desired products. The most often used methods include high-pressure homogenization, hydrodynamic cavitation, bead milling, and ball milling. High-pressure homogenization operates on a basic premise and can cause significant cellular disruption in the biomass.

4.3.1.7 Freezing/Thawing Pretreatment

The freezing and thawing pretreatment technique has a simple mechanism: gradual freezing at temperatures as low as 10 °C below the freezing point of water. This stimulates the formation of ice crystals, which mechanically tear down cell walls [71].

4.3.2 Fermentation

Microalgae-based sugars are fermented using one of two methods: separate hydrolysis and fermentation (SHF) or simultaneous saccharification and fermentation. The sugar solution obtained in the preceding procedures is delivered to the fermenter via a pump. In addition to the glucose solution, S. cerevisiae is introduced to the fermenter to start fermentation. As a result of fermentation, the product is sent to the phase separator and distillation tower via streams. The 99% pure ethanol is then transferred to the storage tank via another stream. As a byproduct, the emitted CO_2 gas is transported to the microalgae culture system via stream and used in microalgae cultivation. Sugars that cannot be converted into bioethanol and other wastes are kept in a tank for use in various operations. The discharged process water is collected and stored for subsequent utilization. Many studies have been conducted over the last 20 years using the SHF approach for microalgal bioethanol production. Nguyen et al. [38] obtained 0.292 g-ethanol/g-biomass from microalgal biomass. Ho et al. [16] used *Zymomonas mobilis* sp. bacteria to extract bioethanol from *Scenedesmus obliquus* microalgae, yielding 0.213 g-ethanol/g-biomass.

4.4 Distillation and Purification

4.4.1 Distillation

The distillation procedure involves boiling the water and ethanol combination. Ethanol has a lower boiling point (78.3 °C) than water (100 °C). As a result, the ethanol evaporates before the water condenses and separates. The eluted sample was tested to confirm the presence of ethanol using various methods. In the Lucas test, the observable creation of turbidity after immediately adding Lucas reagent to the distillate ethanol indicated a positive result for primary alcohol. The elimination of brown color after adding a little amount of iodine solution to the generated ethanol indicated a positive result for the Iodoform Test. In the sodium hydrogen carbonate test, the bioethanol produced effervescence, which was positive (Fig. 4.4).

4.4.2 Purification

Extractive distillation can be performed using ethylene glycol with or without salts such as $CaCl_2$ as an entrainer [4], but other ecologically acceptable entrainers should be investigated. Recently, Ravagnani et al. [54] examined the energy consumption of extractive distillation for ethanol purification using ethylene glycol and tetraethylene glycol as entrainers. They selected tetraethylene glycol since it is likely that ethylene glycol will be

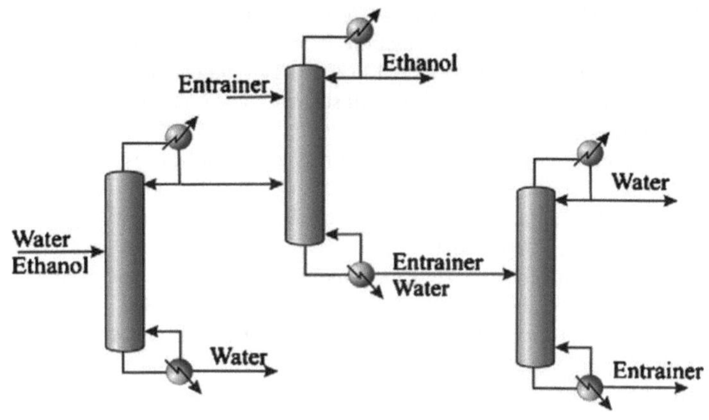

Fig. 4.4 Distillation sequence for purification of bioethanol. *Source* Pacheco-Basulto [42]

banned owing to its toxicity. The results, however, show that using tetraethylene glycol as an entrainer increases energy usage.

Pacheco-Basulto [42] investigates the manufacture of high purity bioethanol from ethylene glycol and ionic liquids (1-butyl-3-methylimidazolium bromide) using simulation and an experimental batch distillation column. It is important to note that ionic liquids (ILs) are considered green solvents with advantages over classical solvents such as low vapor pressure, low toxicity, and high decomposition temperature; however, their cost is significantly higher than that of classical solvents such as benzene, dodecane, and others.

References

1. Aghilinategh M, Barati M, Hamadanian M (2020) The modified supercritical media for one-pot biodiesel production from *Chlorella vulgaris* using photochemically-synthetized $SrTiO_3$ nanocatalyst. Renewable Energy, 160, 176–184. https://doi.org/10.1016/j.renene.2020.06.081.
2. Bhuana DS, Qadariyah L, Panjaitan R, Mahfud M (2020) Optimization of biodiesel production from *Chlorella* sp through in-situ microwave-assisted acid-catalyzed trans-esterification. In IOP Conference Series: Materials Science and Engineering (Vol. 732, No. 1, p. 012004). IOP Publishing. https://doi.org/10.1088/1757-899X/732/1/012004 012004.
3. Brennan L, Owende P (2010) Biofuels from microalgae-a review of technologies for production, processing, and extractions of biofuels and co-products. Renewable and sustainable energy reviews, 14(2), 557–577.
4. Cheirsilp B, Louhasakul Y (2013) Industrial wastes as a promising renewable source for production of microbial lipid and direct transesterification of the lipid into biodiesel. Bioresource technology, 142, 329–337.
5. Chen CY, Zhao XQ, Yen HW, Ho SH, Cheng CL, Lee DJ, Bai FW, Chang JS (2013) Microalgae-based carbohydrates for biofuel production. Biochemical Engineering Journal, 78, 1–10.

6. Chen WH, Huang MY, Chang JS, Chen CY (2014) Thermal decomposition dynamics and severity of microalgae residues in torrefaction. Bioresource Technology, 169, 258–264.
7. Cheng J, Huang R, Li T, Zhou J, Cen K (2015) Physicochemical characterization of wet microalgal cells disrupted with instant catapult steam explosion for lipid extraction. Bioresource technology, 191, 66–72.
8. Chisti Y (2007) Biodiesel from microalgae. Biotechnology advances, *25*(3), 294–306.
9. Das V, Tripathi AM, Borah MJ, Dunford NT, Deka D (2020) Cobalt-doped CaO catalyst synthesized and applied for algal biodiesel production. Renewable Energy, 161, 1110–1119. https://doi.org/10.1016/j.renene.2020.07.040.
10. de Farias Silva CE, Bertucco A (2016) Bioethanol from microalgae and cyanobacteria: a review and technological outlook. Process Biochemistry, 51(11), 1833–1842.
11. Demuez M, Mahdy A, Tomás-Pejó E, González-Fernández C, Ballesteros M (2015) Enzymatic cell disruption of microalgae biomass in biorefinery processes. Biotechnology and Bioengineering, 112(10), 1955–1966.
12. Ghedini E, Taghavi S, Menegazzo F, Signoretto M (2021) A review on the efficient catalysts for algae transesterification to biodiesel. Sustainability, 13(18), 10479.
13. Goettel M, Eing C, Gusbeth C, Straessner R, Frey W (2013) Pulsed electric field assisted extraction of intracellular valuables from microalgae. Algal Research, 2(4), 401–408.
14. Harun R, Yip JW, Thiruvenkadam S, Ghani WA, Cherrington T, Danquah MK (2014) Algal biomass conversion to bioethanol–a step-by-step assessment. Biotechnology journal, 9(1), 73–86.
15. Ho SH, Huang SW, Chen CY, Hasunuma T, Kondo A, Chang, JS (2013a) Bioethanol production using carbohydrate-rich microalgae biomass as feedstock. Bioresource technology, 135, 191–198.
16. Ho SH, Li PJ, Liu CC, Chang JS (2013b) Bioprocess development on microalgae-based CO_2 fixation and bioethanol production using *Scenedesmus obliquus* CNW-N. Bioresource Technology, 145, 142–149.
17. Huang J, Wei M, Ren R, Li H, Liu S, Yang D (2017) Morphological changes of blocklets during the gelatinization process of *Tapioca* starch. Carbohydrate polymers, 163, 324–329.
18. Issariyakul T, Dalai AK (2012) Comparative kinetics of transesterification for biodiesel production from palm oil and mustard oil. The Canadian journal of chemical engineering, 90(2), 342–350.
19. Jung J, Hong SJ, Kim HB, Kim G, Lee M, Shin S, Lee S, Kim DJ, Lee CG, Park Y (2018) Label-free non-invasive quantitative measurement of lipid contents in individual microalgal cells using refractive index tomography. Scientific reports, 8(1), 6524. https://doi.org/10.1038/s41598-018-24393-0.
20. Karpagam R, Rani K, Ashokkumar B, Moorthy IG, Dhakshinamoorthy A, Varalakshmi P (2020) Green energy from *Coelastrella* sp. M-60: Bio-nanoparticles mediated whole biomass transesterification for biodiesel production. Fuel, 279, 118490. https://doi.org/10.1016/j.fuel.2020.118490.
21. Knothe G (2005) Dependence of biodiesel fuel properties on the structure of fatty acid alkyl esters. Fuel processing technology, 86(10), 1059–1070.
22. Kröger M, Müller-Langer F (2012) Review on possible algal-biofuel production processes. Biofuels, 3(3), 333–349.
23. Kuakpetoon D, Wang YJ (2007) Internal structure and physicochemical properties of corn starches as revealed by chemical surface gelatinization. Carbohydrate Research, 342(15), 2253–2263.

24. Lawan I, Garba ZN, Zhou W, Zhang M, Yuan Z (2020) Synergies between the microwave reactor and CaO/zeolite catalyst in waste lard biodiesel production. Renewable Energy, 145, 2550–2560. https://doi.org/10.1016/j.renene.2019.08.008.
25. Lee OK, Kim YH, Na JG, Oh Y K, Lee EY (2013) Highly efficient extraction and lipase-catalyzed transesterification of triglycerides from *Chlorella* sp. KR-1 for production of biodiesel. Bioresource technology, 147, 240–245.
26. Lee SY, Khoiroh I, Vo DVN, Senthil Kumar P, Show PL (2021) Techniques of lipid extraction from microalgae for biofuel production: a review. Environmental Chemistry Letters, 19, 231–251.
27. Liang K, Zhang Q, Cong W (2012) Enzyme-assisted aqueous extraction of lipid from microalgae. Journal of Agricultural and Food Chemistry, 60(47), 11771–11776.
28. López EN, Medina AR, Moreno PAG, Cerdán LE, Valverde LM, Grima EM (2016) Biodiesel production from *Nannochloropsis gaditana* lipids through transesterification catalyzed by *Rhizopus oryzae* lipase. Bioresource technology, 203, 236–244.
29. Luo J, Fang Z, Smith Jr RL (2014) Ultrasound-enhanced conversion of biomass to biofuels. Progress in Energy and Combustion Science, 41, 56–93.
30. Mata TM, Martins AA, Caetano NS (2010) Microalgae for biodiesel production and other applications: a review. Renewable and sustainable energy reviews, 14(1), 217–232.
31. Mata-Alvarez J, Macé S, Llabrés P (2000) Anaerobic digestion of organic solid wastes. An overview of research achievements and perspectives. Bioresource technology, 74(1), 3–16.
32. Miranda JR, Passarinho PC, Gouveia L (2012) Pre-treatment optimization of *Scenedesmus obliquus* microalga for bioethanol production. Bioresource technology, 104, 342–348.
33. Mohamad NR, Marzuki NHC, Buang NA, Huyop F, Wahab RA (2015) An overview of technologies for immobilization of enzymes and surface analysis techniques for immobilized enzymes. Biotechnology & Biotechnological Equipment, 29(2), 205–220.
34. Mubarak M, Shaija A, Suchithra TV (2015) A review on the extraction of lipid from microalgae for biodiesel production. Algal research, 7, 117–123. https://doi.org/10.1016/j.algal.2014.10.008.
35. Mukherjee J, Gupta MN (2016) Dual bioimprinting of *Thermomyces lanuginosus* lipase for synthesis of biodiesel. Biotechnology Reports, 10, 38–43.
36. Naghdi FG, Bai X, Thomas-Hall SR, Sharma K, Schenk PM (2016) Lipid extraction from wet *Chaetoceros muelleri* culture and evaluation of remaining defatted biomass. Algal research, 20, 205–212.
37. Nguyen HC, Nguyen ML, Wang FM, Juan HY, Su CH (2020) Biodiesel production by direct transesterification of wet spent coffee grounds using switchable solvent as a catalyst and solvent. Bioresource technology, 296, 122334. https://doi.org/10.1016/j.biortech.2019.122334.
38. Nguyen MT, Choi SP, Lee JW, Lee JH, Sim SJ (2009) Hydrothermal acid pretreatment of *Chlamydomonas reinhardtii* biomass for ethanol production. Journal of microbiology and biotechnology, 19(2), 161–166.
39. Nielsen PM, Brask J, Fjerbaek L (2008) Enzymatic biodiesel production: technical and economical considerations. European Journal of Lipid Science and Technology, 110(8), 692–700.
40. Osman AI, Mehta N, Elgarahy AM, Al-Hinai A, Al-Muhtaseb AAH, Rooney DW (2021) Conversion of biomass to biofuels and life cycle assessment: a review. Environmental chemistry letters, 19, 4075–4118.
41. Özçimen D, Koçer AT, İnan B, Özer T (2020) Bioethanol production from microalgae. In Handbook of microalgae-based processes and products (pp. 373–389). Academic Press.
42. Pacheco-Basulto JÁ, Hernández-McConville D, Barroso-Muñoz FO, Hernández S, Segovia-Hernández JG, Castro-Montoya AJ, Bonilla-Petriciolet A (2012) Purification of bioethanol

using extractive batch distillation: Simulation and experimental studies. Chemical Engineering and Processing: Process Intensification, 61, 30–35.
43. Pan Y, Alam MA, Wang Z, Huang D, Hu K, Chen H, Yuan Z (2017) One-step production of biodiesel from wet and unbroken microalgae biomass using deep eutectic solvent. Bioresource technology, 238, 157–163.
44. Patil PD, Dandamudi KPR, Wang J, Deng Q, Deng S (2018) Extraction of bio-oils from algae with supercritical carbon dioxide and co-solvents. The Journal of Supercritical Fluids, 135, 60–68.
45. Patil PD, Gude VG, Mannarswamy A, Cooke P, Nirmalakhandan N, Lammers P, Deng S (2012) Comparison of direct transesterification of algal biomass under supercritical methanol and microwave irradiation conditions. Fuel, 97, 822–831.
46. Phukan MM, Chutia RS, Konwar BK, Kataki R (2011) Microalgae *Chlorella* as a potential bio-energy feedstock. Applied Energy, 88(10), 3307–3312.
47. Piligaev AV, Sorokina KN, Samoylova YV, Parmon VN (2018) Lipid production by microalga *Micractinium* sp. IC-76 in a flat panel photobioreactor and its transesterification with cross-linked enzyme aggregates of Burkholderia cepacia lipase. Energy Conversion and Management, 156, 1–9. https://doi.org/10.1016/j.enconman.2017.10.086.
48. Qari H, Rehan M, Nizami AS (2017) Key issues in microalgae biofuels: a short review. Energy Procedia, 142, 898–903.
49. Raheem A, Prinsen P, Vuppaladadiyam AK, Zhao M, Luque R (2018) A review on sustainable microalgae based biofuel and bioenergy production: Recent developments. Journal of cleaner production, 181, 42–59.
50. Ramachandran K, Suganya T, Gandhi NN, Renganathan S (2013) Recent developments for biodiesel production by ultrasonic assist transesterification using different heterogeneous catalyst: A review. Renewable and Sustainable Energy Reviews, 22, 410–418.
51. Ramluckan K, Moodley KG, Bux F (2014) An evaluation of the efficacy of using selected solvents for the extraction of lipids from algal biomass by the soxhlet extraction method. Fuel, 116, 103–108.
52. Ranjan A, Patil C, Moholkar VS (2010) Mechanistic assessment of microalgal lipid extraction. Industrial & Engineering Chemistry Research, 49(6), 2979–2985. https://doi.org/10.1021/ie9016557.
53. Rathnam VM, Modak JM, Madras G (2020) Non-catalytic transesterification of dry microalgae to fatty acid ethyl esters using supercritical ethanol and ethyl acetate. Fuel, 275, 117998. https://doi.org/10.1016/j.fuel.2020.117998.
54. Ravagnani MASS, Reis MHM, Maciel Filho R, Wolf-Maciel MR (2010) Anhydrous ethanol production by extractive distillation: A solvent case study. Process Safety and Environmental Protection, 88(1), 67–73.
55. Raven RP, Gregersen KH (2007) Biogas plants in Denmark: successes and setbacks. Renewable and sustainable energy reviews, 11(1), 116–132.
56. Rendón Castrillón LJ, Ramírez Carmona ME, Vélez Salazar Y (2015) Microalgas para la industria alimenticia. Editorial Universidad Pontificia Bolivariana. Colombia.
57. Rodrigues J, Canet A, Rivera I, Osório NM, Sandoval G, Valero F, Ferreira-Dias S (2016) Biodiesel production from crude *Jatropha* oil catalyzed by non-commercial immobilized heterologous *Rhizopus oryzae* and *Carica papaya* lipases. Bioresource technology, 213, 88–95.
58. Ruiz HA, Rodríguez-Jasso RM, Aguedo M, Kádár Z (2015) Hydrothermal pretreatments of macroalgal biomass for biorefineries. Algal Biorefineries: Volume 2: Products and Refinery Design, 467–491.
59. Sánchez Mirón A, Garcıa MCC, Gómez AC, Camacho FG, Grima EM, Chisti Y (2003) Shear stress tolerance and biochemical characterization of *Phaeodactylum tricornutum* in quasi

steady-state continuous culture in outdoor photobioreactors. Biochemical Engineering Journal, 16(3), 287–297.
60. Sánchez DA, Tonetto GM, Ferreira ML (2018) *Burkholderia cepacia* lipase: A versatile catalyst in synthesis reactions. Biotechnology and bioengineering, 115(1), 6–24.
61. Sarker SD, Nahar L (2012) An introduction to natural products isolation. Natural products isolation, 864, 1–25.
62. Scholz MJ, Weiss TL, Jinkerson RE, Jing J, Roth R, Goodenough U, Posewitz MC, Gerken HG (2014) Ultrastructure and composition of the *Nannochloropsis gaditana* cell wall. Eukaryotic cell, 13(11), 1450–1464.
63. Srivastava G, Paul AK, Goud VV (2018) Optimization of non-catalytic transesterification of microalgae oil to biodiesel under supercritical methanol condition. Energy Conversion and Management, 156, 269–278. https://doi.org/10.1016/j.enconman.2017.10.093.
64. Srivastava RK, Shetti NP, Reddy KR, Kwon EE, Nadagouda MN, Aminabhavi TM (2021) Biomass utilization and production of biofuels from carbon neutral materials. Environmental Pollution, 276, 116731.
65. Teo CL, Idris A (2014) Enhancing the various solvent extraction method via microwave irradiation for extraction of lipids from marine microalgae in biodiesel production. Bioresource Technology, 171, 477–481. https://doi.org/10.1016/j.biortech.2014.08.024.
66. Veeranan T, Kasirajaan R, Gurunathan B, Sahadevan R (2018) A novel approach for extraction of algal oil from marine macroalgae *Ulva fasciata*. Renewable Energy, 127, 64–73.
67. Velazquez-Lucio J, Rodríguez-Jasso RM, Colla LM, Sáenz-Galindo A, Cervantes-Cisneros DE, Aguilar CN, Fernandes BD, Ruiz HA (2018) Microalgal biomass pretreatment for bioethanol production: a review. Biofuel Research Journal, 5(1), 780–791.
68. Vorobiev E, Lebovka N (2015) Selective extraction from food plants and residues by pulsed electric field. Green extraction of natural products: Theory and practice, 307–332.
69. Wadood A, Rana A, Basheer C, Razzaq SA, Farooq W (2020) In situ transesterification of microalgae *Parachlorella kessleri* biomass using sulfonated rice husk solid catalyst at room temperature. BioEnergy Research, 13, 530–541. https://doi.org/10.1007/s12155-019-10060-3.
70. Wahidin S, Idris A, Yusof NM, Kamis NHH, Shaleh SRM (2018) Optimization of the ionic liquid-microwave assisted one-step biodiesel production process from wet microalgal biomass. Energy Conversion and Management, 171, 1397–1404. https://doi.org/10.1016/j.enconman.2018.06.083.
71. Yang C, Liu W, He Z, Thangavel S, Wang L, Zhou A, Wang A (2015) Freezing/thawing pretreatment coupled with biological process of thermophilic *Geobacillus* sp. G1: acceleration on waste activated sludge hydrolysis and acidification. Bioresource Technology, 175, 509–516.
72. Yang HY, Lu WJ, Chen YC, Chen KT, Teng JC, Wan HP (2017) New algal lipid extraction procedure using an amphiphilic amine solvent and ionic liquid. Biomass and Bioenergy, 100, 108–115.
73. Zanuso E, Gomes DG, Ruiz HA, Teixeira JA, Domingues L (2021) Enzyme immobilization as a strategy towards efficient and sustainable lignocellulosic biomass conversion into chemicals and biofuels: current status and perspectives. Sustainable Energy & Fuels, 5(17), 4233–4247.
74. Zhao G, Chen X, Wang L, Zhou S, Feng H, Chen WN, Lau R (2013) Ultrasound assisted extraction of carbohydrates from microalgae as feedstock for yeast fermentation. Bioresource Technology, 128, 337–344.

Integrated Biorefineries and Sustainability 5

Abstract

This chapter examines the future of microalgal bioenergy by analysing integrated biorefinery concepts, advanced photobioreactor design, and the roles of genetics and metabolic engineering. This study explores the potential of algal biofilm and photobioreactor design to improve biomass productivity, while also examining the contributions of genomics and metabolic engineering in optimising microalgal strains for bioenergy applications. This chapter examines the challenges and opportunities associated with microalgal bioenergy, detailing prospective trends and directions for sustainable and efficient bioenergy production. This chapter examines the integration of advanced technologies and strategies, highlighting the potential of microalgal bioenergy to enhance sustainability in energy production.

5.1 Integrated Biorefinery Concepts

Microalgae-derived bioproducts are gaining attention due to their potential, unique properties, present population demands, and escalating costs of increasingly diminishing energy resources. Products made from algae are recognised as sustainable, clean energy and food sources.

The use of biomass as an energy source is constrained by its availability and the maximum conversion efficiencies. According to Basu [3], seasonality, regional dispersion, and relatively low calorific value are barriers to the efficient use of biomass as a renewable carbon source. The sustainable generation of biomass addresses the critical issue of utilizing arable land for food agriculture [15]. To optimize biomass utilization, including waste produced from diverse conversion paths, and transform it into valuable bio-based product streams, the integrated biorefinery is proposed. According to Cherubini

H. N. Akolkar et al., *Bioenergy from Microalgal Biomass*, Synthesis Lectures on Renewable Energy Technologies, https://doi.org/10.1007/978-3-032-03434-2_5

[7] and Ferreira [14], biorefineries are infrastructure facilities that integrate biochemical, combustion, thermochemical, and microorganism growth platforms to efficiently produce sustainable bio-based product streams. These streams include bioenergy, biochemicals, biofuels, and other high-valued bio-products.

5.1.1 Types of Biorefinery

The literature has distinguished three types of biorefinery [13, 21, 40].

a. Phase I biorefinery
b. Phase II biorefinery
c. Phase III biorefinery.

a. **Phase I biorefinery**

Phase I biorefineries have a single major product, a single feedstock, and fixed processing capabilities (single process). They are currently operational and have demonstrated economic viability. Vegetable oil, predominantly rapeseed oil, serves as a feedstock, yielding fixed amounts of biodiesel and glycerin through a process termed transesterification. Consequently, they possess less flexibility to recuperate investment and operational expenses.

b. **Phase II biorefinery**

Phase II biorefineries can only handle one feedstock at a time. Due to their ability to produce a variety of end products (chemicals, energy, and materials), they are able to adjust to changing market demands, price, contractual requirements, and facility operational limitations.

c. **Phase III biorefinery**

The most sophisticated biorefinery technology is known as a Phase III biorefinery. They may produce a diverse array of energy and chemical products (phase II biorefineries) and utilize multiple feedstocks and processing technologies to generate the multitude of industrial products demanded by society. The product diversity affords significant flexibility to adapt to evolving market demands, as a present by-product may transform into a principal product in the future, and offers phase III biorefineries multiple avenues to attain profitability and optimize returns. Moreover, its 'multiple feedstock' characteristic ensures feedstock availability and provides these highly integrated biorefineries the opportunity to choose the most lucrative mix of raw materials [12].

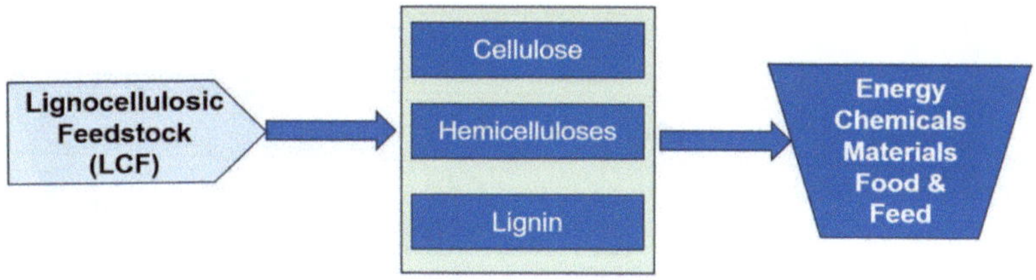

Fig. 5.1 Streamlined schematic representation of a Lignocellulosic feedstock biorefinery

At present, four phase III biorefinery systems are under investigation and development.

- Lignocellulosic feedstock biorefinery
- Whole crop biorefinery
- Green biorefinery
- Two-platform concept biorefinery.

- **Lignocellulosic feedstock biorefinery**

A lignocellulosic biomass biorefinery will typically utilize 'nature-dry' feedstock, including wood, straw, and corn stover. The lignocellulosic feedstock, predominantly composed of polysaccharides and lignin, will undergo fractionation and conversion into various energy and chemical products within the biorefinery through a series of operations. Lignocellulosic biomass comprises three principal polymers: lignin (10–25 wt%) hemicelluloses (20–35 wt%), and cellulose (35–50 wt%), [24] (Fig. 5.1).

- **Whole crop biorefinery**

Cereals (such as maize, wheat, rape, etc.) will be used in a whole crop biorefinery, which will transform the complete plant—straw and grain—into materials, energy and chemicals (Fig. 5.2). The initial phase involves detaching the seed from the straw, with collection occurring concurrently to optimize energy use and labor expenses. After that, the seeds can be processed to yield starch and a variety of goods, such as ethanol and bioplastics (such polylactic acid). As previously mentioned, a lignocellulosic feedstock biorefinery can use a variety of conversion techniques to turn the straw into products.

- **Green biorefinery**

Using a variety of processes, including fermentation, it transforms "natural wet" green biomass—such as green grass, immature cereals, lucerne, clover, algae, etc.—into valuable goods including energy, materials, chemicals, and feed (Fig. 5.3). According to [1], green biomass is usually separated into a nutrient-rich green juice and a press cake that is

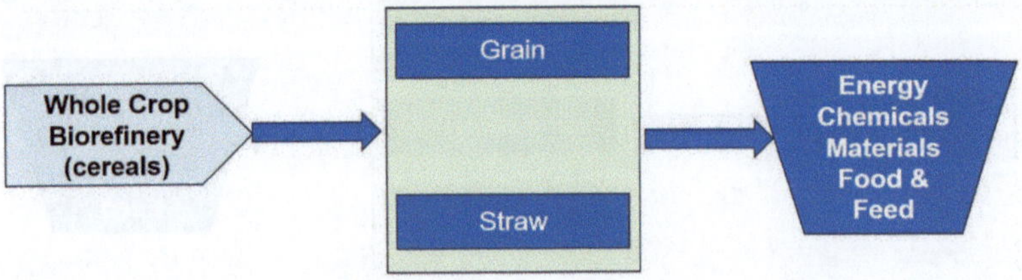

Fig. 5.2 Streamlined schematic representation of a whole crop biorefinery

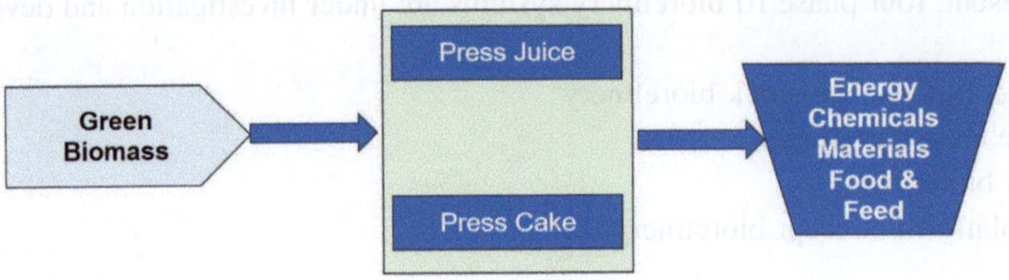

Fig. 5.3 Streamlined schematic representation of a green biorefinery

high in fiber. The green juice comprises various beneficial compounds, including amino acids, organic acids, and pigments. Pasteurized press cake has multiple potential applications, including animal feed, biocomposites, insulating materials, building panels, and energy production.

- **Two-platform concept biorefinery**

Two distinct platforms, one biochemically focused on sugar and the other thermochemically oriented, are used to separate the feedstock, as shown in Fig. 5.4. Both systems can provide energy, chemicals, materials, and even food and feed, thereby utilizing the complete feedstock(s). Biochemical conversion activities constitute the basis of the "sugar platform," which has as its primary focus converting biomass feedstocks into sugars through fermentation. According to Wright and Brown [44], the "syngas platform" can convert biomass into gaseous or liquid intermediate compounds through thermolysis. These chemicals can then be refined into transportation fuels, commodity chemicals, and specialty chemicals. The sugar platform, along with various other non-thermochemical processes expected to be integrated into a biorefinery, is likely to generate waste products that may pose significant challenges in converting them into value-added minerals and chemicals.

According to Ragauskas et al. [33], these byproducts and wastes can be thermochemically converted into energy, which is a valuable asset for biorefineries.

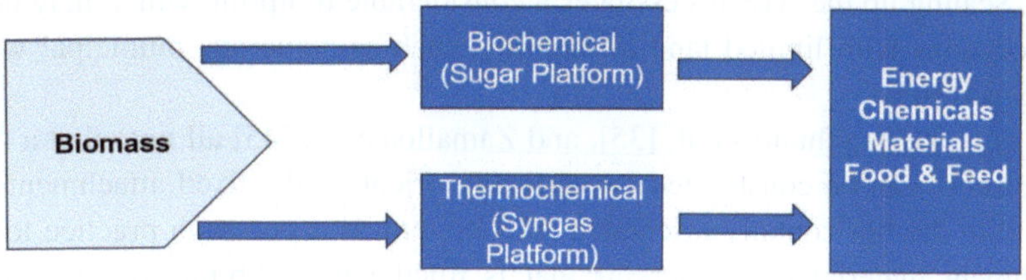

Fig. 5.4 Streamlined schematic representation of a two-platform concept biorefinery

5.2 Algal Biofilm

As a viable substrate for algal production and its subsequent uses, such as wastewater cleanup, biofilm-based algal cultivation has garnered a lot of interest. Methods for algal biofilm cultivation offer a good substitute for suspension-based methods, which are still not commercially feasible. One significant benefit of algal biofilm systems is the simplicity of algae collection via scraping, which eliminates the need for costly harvesting methods like flocculation and centrifugation that are employed in suspension-based systems.

Gross et al. [17] noted that numerous algal biofilm systems have been developed recently, each with its own unique architecture and capacity for biomass generation.

The system's geometric combinations, attachment materials, and design reasoning have evolved greatly depending on its purpose. The biomass present in these systems is readily harvestable. Because of the ideal conditions for photosynthesis, the majority of cells in a biofilm are algae. However, research has shown that the presence of organic carbon in wastewater can contaminate algal biofilms to a certain extent [20, 31]. The geometry of algal biofilm systems exhibits significant variability, ranging from flat plates to rotating drums. Kesaano et al. [22] included cotton, Schnurr et al. [35] lists glass, Blancken et al. [4] listed stainless steel, and Posadas et al. [32] listed polyvinyl chloride as materials used for cell attachment. Sorting algal biofilm systems into stationary and rotating biofilms according to the degree to which the biofilms move through the liquid media.

5.2.1 Stationary Biofilm Design

Biofilms can be laid down either horizontally or vertically, and in the stationary biofilm system, liquid is pumped or otherwise moved over their surfaces. In a typical stationary biofilm system, such as an algal turf scrubber, an attachment sheet—often polyethylene—is laid down horizontally and is flat or slightly angled.

The algal turf scrubber ATS design incorporates minimal moving components, resulting in reduced capital investment, lower energy consumption, and decreased operational expenses. Scaling up the ATS necessitates a considerable footprint, which may not be feasible in locations with limited land availability, such as numerous municipal wastewater treatment facilities.

Genin et al. [16], Schnurr et al. [35], and Zamalloa et al. [45] all note that a "flat-plate biofilm reactor" can be constructed by vertically orienting the fixed attachment material, as opposed to the horizontally orientated ATS system. It is common practice to immerse the attachment material in a reservoir that is filled with bubbling air or a liquid that is pumped about. The flat-plate photobioreactor that uses suspensions is similar to this biofilm reactor; however, this one has an interior attachment substance that helps the biofilm grow.

The algal biofilm's vertical arrangement significantly lessens the land footprint. The ability to completely enclose the liquid reservoir in the flat-plate biofilm reactor further enhances its ability to minimise contamination. Nevertheless, this system's capital expenses would rise due to the clear liquid reservoir. Operating costs are further raised by the additional energy required for air bubbling or liquid circulation.

5.2.2 Rotating Biofilm Design

As an additional example of a biofilm-based system, there is the rotating algal biofilm. The attachment material within the RAB system undergoes alternating rotation between the air phase and the liquid medium. Scraping the biofilm during the air phase is a viable method for harvesting algal biomass in the RAB system [17]. According to Blancken et al. [4], Gross et al. [17], and Christenson and Sims [11], the attachment materials of the RAB system can be cylindrical drums rotated about their centre axis, flat rotating discs, or conveyor belts operated by several shafts.

5.3 Photobioreactor Design

One type of bioreactor that is tailored for growing microalgae is the photobioreactor, or PBR. Microalgal biomass and microalgal product production, wastewater treatment, and carbon dioxide sequestration are just a few of the many uses for these systems [18, 38]. According to Hoh et al. [19], biofilm-based photobioreactors (PBRs) use microalgal biomass that is immobilised on a substrate, which effectively separates the bulk of the biomass from the culture media.

When contrasted with PBRs that rely on suspension, biofilm-based systems provide numerous benefits. Improving operational and harvesting energy efficiency relies heavily on the biofilm's high dry biomass density. Reason being, dewatering is not needed or only partially needed because of the high biomass concentration, which is a prerequisite for downstream processing [29, 30]. According to Olivier et al. [29] and Podola et al. [30], the amount of reactorand/or medium volume needed to grow the same amount of biomass is significantly reduced due to the biofilm's concentrated biomass. Submerged systems and porous substrate photobioreactors (PSBRs) are the two primary forms of biofilm-based photobioreactors [30]. Biofilms are usually immobilised or bonded to impermeable substrates in submerged biofilm photobioreactors (PBRs), which are submerged in the culture media either continuously or intermittently [29, 30].

Biofilms in PSBR are able to cling to microporous substrates, creating an efficient barrier between the biofilm and its environment. The biomass is contained to one side of the substrate due to the substrate's lower pore size compared to the micro-algal cells that comprise the biofilm. According to research by Shi et al. [36], Nowack et al. [28], and Naumann et al. [26], this setup physically isolates the biomass from the medium flow. Microalgal biomass, lipids, pigments, wastewater treatment, and CO_2 sequestration have all been achieved through the use of biofilm-based photobioreactors (PBRs) [19, 29, 30].

All of these uses rely on the high productivity and quick growth of microalgae, which calls for an elevated rate of inorganic carbon fixation.

In contrast to suspension-based PBRs, biofilms do not facilitate the homogeneous mixing of biomass and culture medium. The immobilisation of cells results in mass transfer within dense biofilms being predominantly dependent on diffusion [24, 25, 37]. In comparison to advection, which involves solute transport via liquid flow, mass transfer through diffusion occurs at a significantly slower rate above the micrometre scale [6, 37]. Consequently, gradients of dissolved species, including dissolved inorganic carbon (DIC) concentration, develop within the biofilms. Dense biomass packing is anticipated to result in a light gradient within phototrophic biofilms [23, 37, 43].

5.4 Metabolic Engineering for Microalgal Bioenergy

One of the most encouraging feedstocks for biofuel production is microalgae, and the current state of renewable energy is focused on creating sustainable and alternative energy sources. Microalgae are recognised for their ability to produce significantly higher quantities of biofuels in a reduced timeframe compared to traditional sources derived from plant seeds. Nonetheless, the primary obstacle in transitioning to algae-based biofuel production lies in the numerous complexities associated with microalgal cultivation, as well as the processes of harvesting, concentration, drying, and lipid extraction. Lipids, particularly triacylglycerols (TAGs), which are key building blocks for lipid synthesis, are accumulated by a variety of green microalgae. The analysis of metabolic pathways in the oleaginous microalgae *Chlamydomonas reinhardtii* has revealed novel metabolically significant genes, thereby enhancing lipid production in this organism. The integration of various aspects in metabolic engineering utilising OptFlux, along with effective bioprocess design, provides an interactive overview of the enhancement of lipid production, which ultimately improves oil yield.

Metabolic engineering is the process of manipulating cellular metabolic pathways to induce the production of a specific molecule.

To attain these objectives, several strategies may be employed, as detailed below:

Analysing flux balance, improving photosynthetic efficiency through increased light penetration and decreased cell shading, designing enzymes to aid in lipid biogenesis, pinpointing the rate-limiting enzymes and the specific step involved, capturing and dividing carbon, mathematical modelling, overexpressing a gene or genes, and engineering transcription factors.

The following points delineate the key benefits linked to lipid droplet production in microalgae. Lipid biogenesis comprises three distinct stages: chain elongation, acetyl CoA carboxylation, and the formation of triacylglycerols (TAG). The synthetic biology component requires essential data concerning the organism, particularly microalgae. Reconstructing the metabolic pathways requires whole genome sequencing of microalgae, whether they are models or not. The stoichiometric model, denoted by the equation $S.v = 0$, is used to reconstruct metabolic fluxes. In this equation, v denotes a vector of fluxes, while S represents a matrix constructed by balancing the masses within each cellular compartment of *Chlamydomonas reinhardtii* [5].

5.5 Genetic Engineering of Microalgae for Biorefinery

Genetic engineering involves the manipulation of genes to produce beneficial products. The process of genetic engineering in algae adheres to fundamental steps, which include: (1) Host Selection (2) target gene (3) Transformation (4) Selection and screening. Two steps are particularly critical: transformation techniques and selection [27].

Proteins, pigments, biofuels, nutraceuticals, medications, biohydrogen, and other biological and chemical products are all produced by microalgae and are all categorised as safe products. Their growth rate is relatively rapid, accompanied by significant lipid synthesis. Changing the levels of transcriptional and translational genes can change metabolic pathways. To achieve certain results, genes are either knocked in or knocked out. This method demonstrates greater utility compared to random mutagenesis due to its enhanced accuracy, speed, and precise targeting of the gene. This enhancement boosts productivity and lowers harvesting costs, thereby rendering it more practical for everyday application.

5.6 Challenges in Microalgal Bioenergy

The sustainability and viability of large-scale production remain uncertain despite the many environmental advantages that could be achieved by turning microalgae into biofuels because of the technological and financial difficulties associated with growing, harvesting, drying, and processing microalgal biomass.

The creation of third-generation biofuels from microalgae biomass has drawn more attention throughout the past ten years. Private businesses, academic institutions, governmental bodies, and public research organisations have all undertaken a number of studies centred on the production of biofuel and microalgae. However, as the prices associated with third-generation biofuels are still significantly higher than those of traditional alternatives, a significant amount of time is needed until microalgae-based biofuels are widely adopted commercially [9]. Recent estimates suggest that the cost of microalgae bio-oil ranges from US $300 to $2600. A distinct study has verified that the expense of oil sourced from microalgae is double that of petroleum oil. The existing literature on microalgae-derived biofuel production encompasses numerous studies using Techno-Economic Analysis (TEA) evaluate the feasibility of various technology strategies [41].

Current estimates of biofuel production costs are based on prior research or small-scale experimental investigations. The modelling of these cost estimates often lacks empirical support and validation from actual data. Furthermore, various input assumptions remain to be evaluated, such as lipid productivity, growth rate, nutrient requirements, and energy consumption [8]. According to research, the photobioreactor approach, which has expenses that are roughly two to 2.5 times more than the prior process, has not been studied as much as the production of microalgae in open raceway ponds [2].

Richardson et al. [34] estimate that the current cost range for microalgae-based biofuel is $0.44/L–$8.76/L.

According to the price calculations presented, microalgae oil lacks cost competitiveness relative to other conventional fuel alternatives. Reducing the cost of microalgae biofuels requires improving cultivation methods and streamlining processes.

Furthermore, a reduction in these expenses may be imminent if production and conversion procedures are further systematically optimised [10]. The manufacture of biofuel from microalgae has been the subject of numerous investigations in recent decades. Nonetheless, a restricted number of pilot projects have been implemented in practical settings, and commercialisation has yet to produce significant outcomes [39]. Currently, there is no documented case study on the industrial-scale production of continuous microalgae-based biofuels. Numerous challenges persist in hindering the expansion of production and commercialisation of microalgae-derived biofuels, notably elevated operational and capital investment expenses. Numerous studies have been conducted on microalgae cultivation using both open systems and photobioreactors. The cultivation of microalgae in raceway ponds and photobioreactors poses challenges in biomass propagation. Additional pretreatment procedures and an effective yeast fermentation process are required for the conversion of microalgae-based bioethanol.

Further research into the cultivation of genetically engineered microalgae species with rapid growth traits is necessary to improve existing bioethanol production techniques. The use of bioenergy from microalgae offers significant advantages; however, the widespread adoption of this clean and sustainable energy source faces considerable obstacles.

These challenges include the identification of suitable microalgae strains, the requirements for seasonal variability in feedstock supply, biomass harvesting processes, input resources, and the selection of appropriate conversion methods.

The low rate of sunlight distribution is a significant obstacle that adversely impacts the growth and yield of biomass from microalgae culture. The low biomass concentration is caused in part by the tiny size of microalgae cells. Which in turn increases the cost associated with the harvesting of microalgae biomass. The presence of high moisture content in biomass presents a significant challenge that necessitates the implementation of additional energy-intensive drying stages.

Cultivating microalgae requires a lot more money than traditional farming does because of all the infrastructure that need to be built and maintained to give the microalgae the best possible environment to develop. In response to the difficulties in growing biomass and using techniques for extracting proteins and lipids to make biofuels, some have proposed improving biorefinery processing to cut out the growing and harvesting steps altogether.

Progress in the economic viability of biofuel generation from microalgae biomass has been hindered by the current slow development pace. Successfully overcoming these current challenges is crucial for moving the wider commercial use of biofuel generation from microalgae forward [42].

5.7 Concluding Remarks and Future Prospects

Given the demand for sustainable energy sources, microalgal biofuels have garnered international public and political support. This is because biofuels have the potential to significantly contribute to the fight against energy shortages, fill the gap left by fossil fuels, and lessen the effects of global warming and climate change. Because microalgal growth uses the CO_2 released during the burning of biofuel, the biofuel made from microalgae is typically carbon neutral. Therefore, using biofuels doesn't release more carbon into the environment. Because microalgae can produce a variety of biofuels, including biodiesel, bioethanol, biogas, and biohydrogen, they have a wide range of bioenergy potential. Nevertheless, there are several unanswered concerns regarding the feasibility of large-scale, commercially viable, and environmentally friendly microalgal biofuel production. These include:

(i) Which technology(s) will be most effective for harvesting, microalgal cultivation, and processing to produce biofuels and other valuable by-products?
(ii) Will photosynthetic or heterotropic microorganisms be used for biofuel production?
(iii) How will the CO_2 produced by heterotropic microorganisms be captured?
(iv) Is there a cost-effective way to recycle and reuse CO_2 and different types of nutrients?

The cost gap caused by the high cost of harvesting, cultivation, and processing technologies is the primary obstacle to the generation of biofuel that is economically viable. Both the energy needed for microalgal growing and harvesting and the cost of infrastructure facilities are currently high. Therefore, in order to make the biofuel sector sustainable and profitable, much study is needed. Microalgal research and development (R&D) is still in its early stages, so a significant R&D project is required to: a. Increase the yield of microalgal production through the use of genetic engineering technologies. Gene cloning and manipulation can be made easier with a rudimentary understanding of microalgal biology. Through genetic manipulation, microalgae's biomass and oil content can be increased, as well as their photosynthetic efficiency and capacity to withstand high temperatures.

- Increase the microalgae's metabolic activity to produce effective strains.
- Recognise the physicochemical requirements of cultured water for microalgal species monoculture and/or multiculture.
- Utilise seawater, fermented liquids, agricultural waste, and wastewater from other sources as a fertiliser source; this will improve the environment and reduce the cost of producing biofuel.
- Develop economical technology for harvesting and processing.
- Explore the feasibility of a high-value co-product strategy involving algae in an effort to decrease overall operational costs while improving the yield of biofuel products.

Although there are currently several barriers to the large-scale commercial production of biofuels utilising microalgal feedstock, microalgae have enormous potential for producing biofuels. However, the use of technology and inputs like salt water, CO_2, wastewater, arid terrain, natural light, and aquatic environments are crucial to realising this potential. Since algae have a large potential for creating biofuel and valuable byproducts, both of these possibilities must be investigated effectively and sustainably for the sector to operate successfully.

The issues raised by the use of first- and second-generation biofuels are entirely resolved by microalgae. Using the CO_2 released by thermal power plants (liquid fuels, gas, and coal) to produce biofuel from microalgae could open up new possibilities for reducing greenhouse gas emissions and the effects of climate change. This will also make it possible to replace the costly technology currently in use with inexpensive, renewable carbon capture. Microalgae's ability to trap carbon and co-burn its byproducts with coal would let power producers and customers access reasonably priced electricity.

References

1. Andersen M and Kiel P (2000) Integrated Utilisation of Green Biomass in the Green Biorefinery, Ind. Crops. Prod., 11:129–137.
2. Barry A, Wolfe A, English C, et al (2016) National algal biofuels technology review.
3. Basu P (2018). Biomass Characteristics. Elsevier Inc.
4. Blanken W, Janssen M, Cuaresma M, Libor Z, Bhaiji T, Wiffels R (2014) Biofilm growth of Chlorella sorokiniana in a rotating biological contactor based photobioreactor. Biotechnol Bioeng 111:2436–2445. https://doi.org/10.1002/bit.25301.
5. Boyle NR and Morgan JA (2009) Flux balance analysis of primary metabolism in Chlamydomonas reinhardtii. BMC Syst. Biol. 3:4. https://doi.org/10.1186/1752-0509-3-4.
6. Brenn G (2017) Analytical Solutions for Transport Processes: Fluid Mechanics, Heat and Mass Transfer. Berlin, Heidelberg: Springer, 2017:239–70. https://doi.org/10.1007/978-3-662-51423-8.
7. Cherubini F (2010) The biorefinery concept: Using biomass instead of oil for producing energy and chemicals. Energy Conversion and Management, 51:1412–1421. https://doi.org/10.1016/j.enconman.2010.01.015.
8. Chia SR, Ong HC, Chew KW et al (2018) Sustainable approaches for algae utilisation in bioenergy production. Renew Energy 129:838–852. https://doi.org/10.1016/j.renene.2017.04.001.
9. Chowdhury H, Loganathan B (2019) Third-generation biofuels from microalgae: a review. Curr Opin Green Sustain Chem 20:39–44. https://doi.org/10.1016/j.cogsc.2019.09.003.
10. Chowdhury H, Loganathan B, Mustary I, et al (2019) Algae for biofuels: the third generation of feedstock. In: Second and third generation of feedstocks. Elsevier, 323–344. https://doi.org/10.1016/B978-0-12-815162-4.00012-4.
11. Christenson L, Sims R (2012) Rotating algal biofilm reactor and spool harvester for wastewater treatment with biofuels by-products. Biotechnol Bioeng 109:1674–1684. https://doi.org/10.1002/bit.24451.
12. de Jong E, van Ree R, van Tuil R and Elbersen W (2006) Biorefineries for the Chemical Industry–A Dutch Point of View, in Biorefineries–Biobased Industrial Processes and Products. Status Quo and Future Directions, B. Kamm, M. Kamm, and P. Gruber (Eds), Wiley-VCH.
13. Fernando S, Adhikari S, Chandrapal C and Murali N (2006) Biorefineries: Current Status, Challenges, and Future Direction, Energy and Fuels, 20:1727–1737. https://doi.org/10.1021/ef060097w.
14. Ferreira AF (2017) Biorefinery Concept. In: Biorefineries: Targeting Energy, High Value Products and Waste Valorisation, (Eds.) Springer International Publishing. Cham, 1–20.
15. Gavrilescu M (2014) Biorefinery Systems: An Overview. Bioenergy Research: Advances and Applications, 219–241. https://doi.org/10.1016/B978-0-444-59561-4.00014-0.
16. Genin S, Aitchison S, Allen D (2014) Design of algal film photobioreactors: material surface energy effects on algal film productivity, colonization and lipid content. Bioresour Technol 155:136–143. https://doi.org/10.1016/j.biortech.2013.12.060.
17. Gross M, Jarboe D, Wen Z (2015) Biofilm-based algal cultivation systems, Appl Microbiol Biotechnol. 99:5781–5789. https://doi.org/10.1007/s00253-015-6736-5.
18. Gupta PL, Lee S-M, Choi H-J. A mini review: photobioreactors for large scale algal cultivation. World J Microb Biot 2015;31: 1409–17. https://doi.org/10.1007/s11274-015-1892-4.
19. Hoh D, Watson S, Kan E (2016) Algal biofilm reactors for integrated wastewater treatment and biofuel production: a review. Chem Eng J. 287:466–73. https://doi.org/10.1016/j.cej.2015.11.062.

20. Irving T, Allen D (2011) Species and material considerations in the formation and development of microalgal biofilms. Appl Microbiol Biotechnol 92:283–294. https://doi.org/10.1007/s00253-011-3341-0.
21. Kamm B and Kamm M (2004) Principles of Biorefineries, Appl Microbiol Biotechnol, 64:137–145. https://doi.org/10.1007/s00253-003-1537-7.
22. Kesaano M, Gardner R, Moll K, Lauchnor E, Gerlach R, Petron B, Sims R (2015) Dissolved inorganic carbon enhanced growth, nutrient uptake, and lipid accumulation in wastewater grown microalgal biofilms. Bioresour Technol 180:7–15. https://doi.org/10.1016/j.biortech.2014.12.082.
23. Liehr SK, Wayland Eheart J, Suidan MT (1988) A modeling study of the effect of pH on carbon limited algal biofilms. Water Res 22:1033–41. https://doi.org/10.1016/0043-1354(88)90151-0.
24. Ma J, Shi S, Jia X, Xia F, Ma H, Gao J, Xu J (2019) Advances in catalytic conversion of lignocellulose to chemicals and liquid fuels. Journal of Energy Chemistry, 36:74–86. https://doi.org/10.1016/j.jechem.2019.04.026.
25. Murphy TE, Berberoglu H (2014) Flux balancing of light and nutrients in a biofilm photobioreactor for maximizing photosynthetic productivity. Biotechnol Progress. 30:348–59. https://doi.org/10.1002/btpr.1881.
26. Naumann T, Cebi Z, Podola B, Melkonian M (2013) Growing microalgae as aquaculture feeds on twin-layers: a novel solid-state photobioreactor. J Appl Phycol 25:1413–20. https://doi.org/10.1007/s10811-012-9962-6.
27. Ng IS, Tan SI, Kao PH, Chang YK, Chang JS (2017) Recent Developments on Genetic Engineering of Microalgae for Biofuels and Bio-Based Chemicals. Biotechnol. J. 12. https://doi.org/10.1002/biot.201600644.
28. Nowack ECM, Podola B, Melkonian M (2005) The 96-well twin-layer system: a novel approach in the cultivation of microalgae. Protist 156:239–51. https://doi.org/10.1016/j.protis.2005.04.003.
29. Olivieri G, Salatino P, Marzocchella A (2014) Advances in photobioreactors for intensive microalgal production: configurations, operating strategies and applications. J Chem Technol Biotechnol, 89:178–95. https://doi.org/10.1002/jctb.4218.
30. Podola B, Li T, Melkonian M (2017) Porous substrate bioreactors: a paradigm shift inmicroalgal biotechnology? Trends Biotechnol 35:121–32. https://doi.org/10.1016/j.tibtech.2016.06.004.
31. Pohlon E, Marxsen J, Kusel K (2009) Pioneering bacterial and algal communities and potential extracellular enzyme activities of stream biofilms. FEMS Microb Ecol 71:364–373. https://doi.org/10.1111/j.1574-6941.2009.00817.x.
32. Posadas E, Garcia-Encina P, Soltau A, Dominguez A, Diaz I, Munoz R (2013) Carbon and nutrient removal from centrates and domestic wastewater using algal–bacterial biofilm bioreactors. Bioresour Technol 139:50–58. https://doi.org/10.1016/j.biortech.2013.04.008.
33. Ragauskas AJ, Williams CK, Davison BH, Britovsek G, Cairney J, Eckert CA, Frederick Jr WJ, Hallett JP, Leak DJ, Liotta CL, Mielenz JR, Murphy R, Templer R, Tschaplinski T, (2006) The Path Forward for Biofuels and Biomaterials, Science, 311:484–489. https://doi.org/10.1126/science.1114736.
34. Richardson JW, Johnson MD, Outlaw JL (2012) Economic comparison of open pond raceways to photo bio-reactors for profitable production of algae for transportation fuels in Southwest. Algal Res 1:93–100. https://doi.org/10.1016/j.algal.2012.04.001.
35. Schnurr P, Espie G, Allen G (2013) Algae biofilm growth and the potential to stimulate lipid accumulation through nutrient starvation. Bioresour Technol 136:337–344. https://doi.org/10.1016/j.biortech.2013.03.036.

36. Shi J, Podola B, Melkonian M (2007) Removal of nitrogen and phosphorus from wastewater using microalgae immobilized on twin layers: an experimental study. J Appl Phycol, 19:417–23. https://doi.org/10.1007/s10811-006-9148-1.
37. Siegrist H, Gujer W (1985) Mass transfer mechanisms in a heterotrophic biofilm. Water Res, 19:1369. https://doi.org/10.1016/0043-1354(85)90303-3.
38. Singh RN, Sharma S (2012) Development of suitable photobioreactor for algae production–a review. Renew Sust Energ Rev 16:2347–53. https://doi.org/10.1016/j.rser.2012.01.026.
39. Su Y, Song K, Zhang P et al (2017) Progress of microalgae biofuel's commercialization. Renew Sustain Energy Rev 74:402–411. https://doi.org/10.1016/j.rser.2016.12.078.
40. van Dyne DL, Blase MG and Davis L (1999)Clements, A Strategy for Returning Agriculture and Rural America to Long-Term Full Employment Using Biomass Refineries, in Perspectives on New Crops and New Uses, J. Janeck (Ed.), ASHS Press, Alexandria.
41. Vo Hoang Nhat P, Ngo HH, Guo WS et al (2018) Can algaebased technologies be an affordable green process for biofuel production and wastewater remediation? Bioresour Technol 256:491–501. https://doi.org/10.1016/j.biortech.2018.02.031.
42. Wang J, Yin Y (2018) Fermentative hydrogen production using pretreated microalgal biomass as feedstock. Microb Cell Fact 17:22. https://doi.org/10.1186/s12934-018-0871-5.
43. Wolf G, Picioreanu C, van Loosdrecht MCM (2007) Kinetic modeling of phototrophic biofilms: The PHOBIA model. Biotechnol Bioeng 97: 1064–79. https://doi.org/10.1002/bit.21306.
44. Wright M and Brown RC (2007) Comparative Economics of Biorefineries Based on the Biochemical and Thermochemical Platforms, Biofuels, Bioprod. Bioref., 1: 49–56. https://doi.org/10.1002/bbb.8.
45. Zamalloa C, Boon N, Verstraete W (2013) Decentralized two-stage sewage treatment by chemical-biological flocculation combined with microalgae biofilm for nutrient immobilization in a roof installed parallel plate reactor. Bioresour Technol 130:152–160. https://doi.org/10.1016/j.biortech.2012.11.128.

MIX
Papier aus verantwortungsvollen Quellen
Paper from responsible sources
FSC® C105338

FSC
www.fsc.org

If you have any concerns about our products,
you can contact us on
ProductSafety@springernature.com

In case Publisher is established outside the EU,
the EU authorized representative is:
Springer Nature Customer Service Center GmbH
Europaplatz 3, 69115 Heidelberg, Germany

Printed by Libri Plureos GmbH
in Hamburg, Germany